Mark McCaughrean

111 Places in Space That You Must Not Miss

emons:

We dedicate this book to Bruce McCandless II, the first human to fly untethered in space. The iconic image from NASA's STS-41B in 1984 continues to spark our imaginations here on Earth and to inspire us to believe we can all achieve things that are beyond our wildest dreams.

For Tigger, my beloved fellow astral traveller.   MJM

© Emons Verlag GmbH
Cäcilienstraße 48, 50667 Köln
info@emons-verlag.de
All rights reserved
© Photographs see p. 232
© Cover icon: Bruce McCandless II's 1984 Untethered Space Walk,
STS-41B mission; photo by Robert L. "Hoot" Gibson,
used with permission from the McCandless Family
Design: Eva Kraskes, based on a design
by Lübbeke | Naumann | Thoben
Edited by Karen E. Seiger
Printing and binding: sourc-e GmbH
Printed in Europe 2025
ISBN 978-3-7408-0601-9
First edition

Guidebooks for Locals & Experienced Travelers
Join us in uncovering new places around the world at
www.111places.com

Foreword

This book is a portal that leads from our home world into the wider Universe. It imagines that it's possible to cross vast gulfs of spacetime on journeys to amazing locations in our Solar System, the Milky Way galaxy, and into Deep Space beyond. It imagines that you'd survive the extreme conditions you'll find out there, and that you'll be able to see the invisible and discover a panoply of wonders through magical, multiwavelength goggles.

Some of these locations have been and may yet be explored by humans in person. And relativity tells us that you could, in principle, travel across millions of light years and only age a few years yourself, at least as long as you accept the consequences of doing so.

But for me, it's just as remarkable that we have the curiosity and technology to explore and study those places without needing to go ourselves. We've sent robot avatars close to the Sun, to planets, moons, asteroids, and comets, and to the edge of interstellar space. Beyond that, our telescopes capture light from across the electromagnetic spectrum, allowing us to see new stars and planets forming, the event horizons of black holes, and the birth of the Universe 13.8 billion years ago.

We have learned much about the scale, complexity, and history of the cosmos. We've also learned that we're intimately part of it, made from atoms created in the Big Bang, in stars, and in gigantic cosmic explosions. Those atoms have made minds capable of understanding this idea and of asking further questions about our ephemeral place in the Universe through science, and also art, culture, and philosophy.

As an astronomer, I've been privileged to use some of the largest telescopes on Earth and off it, and to travel to other worlds vicariously through the eyes of robotic spacecraft. In this book, I hope to share some of what my many colleagues and I have found so far and of the mysteries that remain. So pack your goggles, protective gear, and imagination into your spacecraft. Pick a destination … and go!

— MJM

111 Places

1___ Apollo 12 | Solar System
You can look, but please don't touch | 10

2___ Arrokoth | Solar System
Life in the dark beyond Neptune | 12

3___ Caloris Basin | Solar System
The biggest impact crater on Mercury | 14

4___ Ceres | Solar System
The inner Solar System's only dwarf planet | 16

5___ Charon | Solar System
Who pays the ferryman? | 18

6___ The Cliffs of Hathor | Solar System
Close to the edge | 20

7___ Comet 67P/Churyumov-Gerasimenko | Solar System
The adventure of a lifetime | 22

8___ Dimorphos | Solar System
Saving the world, one pile of rocks at a time | 24

9___ Dust Devils of Mars | Solar System
Wandering graffiti artists | 26

10___ Europa | Solar System
An icy world in motion | 28

11___ The "Face" on Mars | Solar System
Much ado about nothing | 30

12___ Far Side of the Moon | Solar System
Not your father's prog rock album | 32

13___ Ganymede | Solar System
Some moons are more equal than others | 34

14___ Hubble Space Telescope | Solar System
Beauty is eternity gazing at itself in a mirror | 36

15___ Hyperion | Solar System
A spongy, cosmic beehive | 38

16___ International Space Station | Solar System
An accessible, if expensive, tourist destination | 40

17___ Io | Solar System
A land of ice and fire | 42

18___ Jupiter | Solar System
King of the planets | 44

19 — Lutetia | Solar System
Heavy metal in the sky | 46

20 — Maat Mons | Solar System
Is Venus still geologically active? | 48

21 — Mars | Solar System
The planet of dreams or nightmares? | 50

22 — Mercury | Solar System
The Solar System's problem child | 52

23 — Miranda | Solar System
A paradise for space geologists and daredevils | 54

24 — Near Side of the Moon | Solar System
A cosmic treasure island? | 56

25 — Neptune | Solar System
Still the most distant planet in the Solar System | 58

26 — North Polar Hexagon | Solar System
Who's a pretty polygon then? | 60

27 — Olympus Mons | Solar System
Atop the tallest volcano in the Solar System | 62

28 — 'Oumuamua | Solar System
Our first known interstellar visitor | 64

29 — The Pale Blue Dot | Solar System
If you lived here, you'd be home now | 66

30 — Phobos | Solar System
A natural space station orbiting the Red Planet | 68

31 — Plumes of Enceladus | Solar System
Water, water, everywhere, nor any drop to drink | 70

32 — Pluto | Solar System
The god of the dark Underworld | 72

33 — Rings of Saturn | Solar System
Not all those who wander are lost | 74

34 — The Sun | Solar System
The star in our back garden | 76

35 — Sunspots | Solar System
Tracers of the Sun's internal cycle | 78

36 — Titan | Solar System
A smoggy moon with marvels below | 80

37 — Triton | Solar System
A smörgåsbord of icy delights | 82

38 — Uranus | Solar System
An enigmatic, sideways world | 84

39 __ Venus | Solar System
   Our deadly planetary neighbor | 86

40 __ AG Carinae | Milky Way
   Getting ready for its grand finale | 88

41 __ Alpha Centauri System | Milky Way
   Our nearest stellar neighbor(s) | 90

42 __ Betelgeuse | Milky Way
   All the colors of the stars | 92

43 __ Boomerang Nebula | Milky Way
   Baby, it's cold outside | 94

44 __ Bubble Nebula | Milky Way
   Grace under pressure | 96

45 __ Carina's Bok Globules | Milky Way
   Islands in the storm | 98

46 __ Cat's Eye Nebula | Milky Way
   The beginning of the end for a star | 100

47 __ Cederblad 110 | Milky Way
   Expect a frosty reception | 102

48 __ Cometary Globule 4 | Milky Way
   Art imitating life? | 104

49 __ Cosmic Bat Nebula | Milky Way
   Appearances can be deceptive | 106

50 __ Crab Nebula | Milky Way
   A millennium-old stellar explosion | 108

51 __ CW Leonis | Milky Way
   Ashes to ashes, dust to dust | 110

52 __ Cygnus X | Milky Way
   The spot for students of star formation | 112

53 __ Elephant's Trunk Nebula | Milky Way
   Reaching for the stars | 114

54 __ Galactic Center | Milky Way
   Into the heart of the beast | 116

55 __ HD209458b | Milky Way
   The first known transiting exoplanet | 118

56 __ Herbig-Haro 212 | Milky Way
   A cosmic double lightsaber | 120

57 __ Horsehead Nebula | Milky Way
   A dusty equine near the hunter's belt | 122

58 __ HR8799 | Milky Way
   An extrasolar orrery | 124

59 — Lagoon Nebula | Milky Way
Cloudy, with a chance of twisters | 126

60 — LL Pegasi | Milky Way
Two stars locked in a dusty death spiral | 128

61 — The Milky Way | Milky Way
Seeing the forest for the trees | 130

62 — NGC1999 | Milky Way
When is a hole not a hole but actually is a hole? | 132

63 — The OMC-1 Explosion | Milky Way
Cosmic shrapnel | 134

64 — Omega Centauri | Milky Way
King of the globular clusters | 136

65 — Orion Nebula | Milky Way
Massive star formation below the belt | 138

66 — Pillars of Creation | Milky Way
Towering columns of gas, dust, and young stars | 140

67 — The Pleiades | Milky Way
The star cluster with a thousand names | 142

68 — Polaris | Milky Way
The North Star … for now at least | 144

69 — R Aquarii | Milky Way
A story of symbiosis between little and large | 146

70 — Ring Nebula | Milky Way
When is a planet not a planet? | 148

71 — RS Puppis | Milky Way
Rhythm of the stars | 150

72 — Serpens Nebula | Milky Way
A hissing nest of star formation | 152

73 — Sirius | Milky Way
Twinkle, twinkle, little star | 154

74 — Taurus-Auriga Clouds | Milky Way
A low-density neighborhood for young stars | 156

75 — Terzan 5 | Milky Way
Digging for ancient galactic fossils | 158

76 — Vela Supernova Remnant | Milky Way
A cosmic memento mori | 160

77 — Westerlund 1 | Milky Way
A giant lurking behind a dark veil | 162

78 — Zeta Ophiuchi | Milky Way
Big star in a hurry | 164

79 __ Andromeda Galaxy | Deep Space
Our ever-closer neighbor | 166

80 __ The Antennae | Deep Space
Magnificent chaos as galaxies collide | 168

81 __ Arp 282 | Deep Space
Caught in the act | 170

82 __ Cartwheel Galaxy | Deep Space
Intergalactic hit and run | 172

83 __ Cigar Galaxy | Deep Space
Bursting with new stars | 174

84 __ The CMB | Deep Space
Left-over glow from the Big Bang | 176

85 __ ESO 137-001 | Deep Space
The pressure's on this high-speed medusa | 178

86 __ ESO 306-17 | Deep Space
The dangers of overconsumption | 180

87 __ Fornax A | Deep Space
Dusty heart of a hybrid galaxy | 182

88 __ The Great Attractor | Deep Space
The inexorable pull of gravity | 184

89 __ Hanny's Voorwerp | Deep Space
The power of crowdsourcing | 186

90 __ JADES Origins Deep Field | Deep Space
The first galaxies in the Universe | 188

91 __ MACS J0025.4-1222 | Deep Space
On the trail of the invisible | 190

92 __ Messier 74 | Deep Space
Design is how it works | 192

93 __ Messier 87 | Deep Space
King of its neighborhood | 194

94 __ Messier 106 | Deep Space
Surveying the Universe | 196

95 __ NGC474 | Deep Space
Shells on a galactic seashore | 198

96 __ NGC660 | Deep Space
What's your inclination? | 200

97 __ NGC1365 | Deep Space
A galaxy walks into a bar … | 202

98 __ NGC2276 | Deep Space
Some galaxies have all the luck | 204

99 — NGC2775 | Deep Space
Pulling the wool over your eyes | 206

100 — NGC4753 | Deep Space
Cosmic filigree and shadow | 208

101 — NGC7331 | Deep Space
The Milky Way's almost twin | 210

102 — Nubecula Major | Deep Space
What's in a name? | 212

103 — Perseus Cluster | Deep Space
A life surfing the cosmic web | 214

104 — Pōwehi | Deep Space
The dark heart of a supermassive black hole | 216

105 — SMACS J0723.3-7327 | Deep Space
Cosmic lens with a presidential seal of approval | 218

106 — Sombrero Galaxy | Deep Space
A strange, dusty ring around a central monster | 220

107 — Spanish Dancer Galaxy | Deep Space
I'm forever blowing bubbles | 222

108 — Spindle Galaxy | Deep Space
O what a tangled web we weave | 224

109 — Stephan's Quintet | Deep Space
All is not what it seems in this group of galaxies | 226

110 — Supernova 1987A | Deep Space
The most recent supernova in our neighborhood | 228

111 — Tarantula Nebula | Deep Space
Shelob's cosmic cousin | 230

SOLAR SYSTEM

1 Apollo 12
You can look, but please don't touch

On some of your journeys, you'll be following in the footsteps of brave explorers who have gone before you. And some of those footsteps are real. They're archaeological treasures worthy of preservation. Nowhere is this more evident than at the six landing sites of the first human missions to the Moon.

The touchdown point of NASA's Apollo 12 in the Ocean of Storms is one such site. If you fly overhead near sunset or sunrise, a long shadow will reveal the location of the descent stage of the lunar module Intrepid to the left of a large crater. It has been standing there since astronauts Pete Conrad (1930–1999) and Al Bean (1932–2018) started their journey home to Earth on November 20, 1969. The US flag, scientific experiments, and other equipment surround it. But look more closely towards the bright upper edge of the crater, and you'll notice another shadow – that's the Surveyor 3 probe which had soft-landed there two years earlier. Apollo 12 was deliberately aimed nearby as a test of precision landing technology.

Around the crater rim, you'll see the faint trail made by Conrad and Bean as they walked the 300 meters (330 yards) or so between Intrepid and Surveyor 3. Their footprints are still visible more than half a century later. Since there's no rain or wind on the Moon, they'll likely last for millions of years, only slowly eroded by micrometeorite impacts. These historical legacies of humankind's first steps on another world deserve to be protected. Indeed, US laws today prohibit any disruption of the Apollo sites by countries and private companies working with NASA.

Ironically, before they returned to the lunar module, Conrad and Bean cut some parts off Surveyor 3, arguably also an archaeological treasure. These parts were returned to Earth to allow engineers to study the effects of long duration exposure on the Moon.

Address Oceanus Procellarum (3.0°S, 336.6°E), The Moon, Orbiting Earth, Third Planet from the Sun, The Solar System | Getting there It takes three days to reach the Moon from Earth, plenty of time to contemplate the storied history of this first route navigated by humans towards the stars. | Tip Some 180 km (113 mi) east is the Apollo 14 landing site. That's where you'd see the first extraterrestrial golf balls, which Alan Shepard (1923–1998) struck during that 1971 mission.

SOLAR SYSTEM

2 Arrokoth

Life in the dark beyond Neptune

In the gloom beyond the orbit of Neptune, the last of the giant planets in our Solar System, innumerable dark, icy bodies await your arrival. Arrokoth is the smallest of these yet seen up close, so you'll be exploring a strange world.

Arrokoth lies in the Kuiper Belt, named for Dutch astronomer Gerard Kuiper (1905–1973). In 1951, following earlier ideas by other astronomers, he speculated on its existence as a remnant of the early Solar System. Today, we know the Kuiper Belt is a doughnut-shaped torus beyond Neptune's orbit, extending to roughly twice that distance, with perhaps another, separate belt further out. It likely contains hundreds of thousands of objects larger than 100 km (62 mi) in diameter and perhaps trillions of smaller comets.

The best known Kuiper Belt object is Pluto, king of the dwarf planets, which the NASA New Horizons mission visited in 2015, along with its moon Charon. The probe didn't stop there. It continued its journey outwards, and on New Year's Day in 2019, it flew past Arrokoth, which means "sky" or "cloud" in the Powhatan language. Retracing those steps, you'll find that Arrokoth comprises two bodies 15 and 21 km (9 and 13 mi) across, one pancake-like, the other walnut shaped. They probably formed independently in the early years of the Solar System, accreting material from their surroundings. They then started to orbit each other while still growing, until they slowly touched and merged.

Descending to the surface, you'll find a very low-gravity world with a composition similar to that of comets, namely a porous, low-density mixture of dust and ices, with a coating of water ice, methanol, hydrogen cyanide, and sulfur. Complex organic molecules called tholins are produced from carbon dioxide, methane, and ethane by cosmic rays and ultraviolet light from the Sun – they give Arrokoth its distinctive red hue, shared by many Kuiper Belt objects.

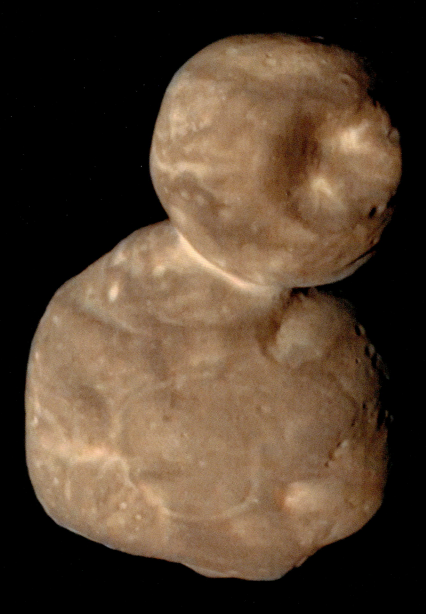

Address The Kuiper Belt, The Solar System | **Getting there** Arrokoth moves slowly across the sky in its 298-year orbit and will be in Sagittarius, the archer, and Capricornus, the horned goat, until at least 2045. Head in that direction for 6.7 billion km (4.2 million mi). | **Tip** 2014 OS393 is close by, another candidate Kuiper Belt object for the New Horizons flyby. It has since been found to comprise two 30 km (19 mi) components separated by a distance of 150 km (93 mi).

3 Caloris Basin
The biggest impact crater on Mercury

As you descend towards the scorching surface below, the vast scale of the Caloris Basin will quickly become apparent. Spanning over 1,500 km (940 mi) across or 10 percent of the circumference of Mercury, this giant crater was created almost four billion years ago when an asteroid at least 100 km (62 mi) in size struck the closest planet to the Sun.

The impact created two broken rings of mountains and cliffs up to 2 km (1.2 mi) high, while the basin itself was flooded with lava escaping from the interior of the planet. In the billions of years since the impact, the region has been struck again by numerous smaller asteroids and meteorites, making other craters. Several of the younger craters have groups of strange, irregular depressions with bright floors and rims known as "hollows," which are probably caused by sublimation of sulfur compounds brought closer to the surface by the more recent events. Now turn on your gas sensors, and you'll discover that Mercury has an extremely thin atmosphere, including hydrogen, helium, and oxygen. There are also sodium, potassium, and calcium, volatile elements that should have been removed by the intense heat of the Sun long ago.

So where are they coming from today? The Caloris Basin is a rich source for these elements, and it's thought that they emerge from the "hollows" in the young craters, as well as from material brought to the surface in the past few hundred million years in explosive volcanic events similar to the one that buried Pompeii.

As you start your journey home, take a look back at Mercury. If the conditions are right, you'll see a giant, yellow tail extending up to 24 million km (15 million mi) from the planet. Atoms in the thin atmosphere are blown into this tail by the strong sunlight, which then causes the sodium in particular to glow, giving it that characteristic streetlight hue.

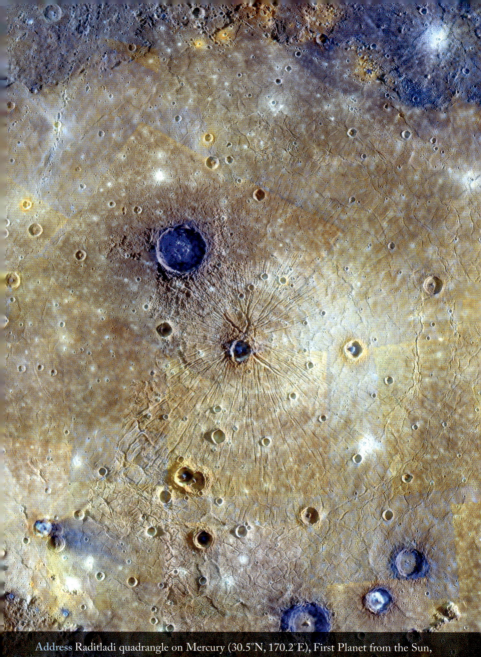

Address Raditladi quadrangle on Mercury (30.5°N, 170.2°E), First Planet from the Sun, The Solar System | **Getting there** Mercury is close to the Sun, and you'll need to shed a lot of speed using flybys at Venus and Mercury itself in order to orbit and land. | **Tip** Visit the antipode of the Caloris Basin to see chaotic terrain likely linked to the impact due either to seismic waves in the planet or ejecta above it converging there (21°S, 341.5°E).

4 Ceres

The inner Solar System's only dwarf planet

Sicilian astronomer and priest Giuseppe Piazzi (1746–1826) discovered Ceres in 1801, and it has had an identity crisis ever since. Almost 1,000 km (620 mi) in diameter, it was first classified as a planet, not least because astronomers expected to find one between Mars and Jupiter. But it soon became clear that Ceres is part of a huge family of objects in the same region: the asteroids. Ceres is the largest asteroid and one of a handful of objects in the inner Solar System thought to have stopped growing before becoming full planets. Partially in recognition of this fact, Ceres was recently relabeled as a "dwarf planet," joining Pluto and other objects beyond Neptune in this new category.

Despite all this to-ing and fro-ing, you'll find Ceres to be a fascinating destination. Beneath a rocky crust, 50 percent of its volume is water, a mix of salty brine and ice perhaps surrounding an inner metal core. It has a tenuous atmosphere of water vapor that may come from exposed surface ice or cryovolcanoes. The most prominent volcano is Ahuna Mons, rising 5 km (3 mi) above the surface. Interestingly, it is antipodal to one of Ceres' largest craters, Kerwan, 1,500 km (930 mi) away on the opposite side. This might not be a coincidence. One idea is that the impact that created Kerwan sent giant seismic waves through Ceres and cracked the surface, allowing salty, muddy cryomagma to well up and build Ahuna Mons.

Another curious feature found on Ceres by NASA's Dawn mission is that some of its youngest craters have bright white patches of salts rich in sodium and aluminum. They're probably the result of brine leaking out when the craters were made, and then evaporating – similar processes may still be happening today. If you were to land on Ceres, perhaps you'd discover more about the origins of this salty little planet and how it manages to remain geologically active.

Address The Asteroid Belt, between Mars and Jupiter, The Solar System | Getting there You need to pick the right moment for your journey – Ceres comes within 240 million km (150 million mi) of Earth at closest approach, but the distance can be more than double that when Earth and Ceres are on opposite sides of the Sun. | Tip A challenging side trip while in the asteroid belt would be to Pallas, a left-over protoplanet like Ceres. It has an inclined orbit and is hard to reach – you could be its first visitor.

5 Charon

Who pays the ferryman?

Once you've made the long journey to the outer reaches of the Solar System to visit Pluto, it'll be hard to miss Charon, its largest moon. Charon is so large, half the diameter of Pluto, that they both orbit a common center of mass that lies outside the body of Pluto. The pair is perhaps better thought of as a binary dwarf planet system. How did Charon and Pluto end up together? Astronomers think that they formed independently in the Kuiper Belt and collided, but slowly enough to avoid destroying each other. After some time stuck to each other, they separated again and remained a pair.

Despite its size, Charon was only discovered by Jim Christy in 1978, almost 50 years after Pluto. Far from the Sun, they're faint and close enough together that they're hard to see separately from a distance, but once you get close, you'll find that Charon is a world unto itself. While Pluto's surface is mostly nitrogen ice, Charon is covered in water ice, with patches of ammonia hydrates. At its north pole is a large depression covered in a red-brown, organic tar known as tholins. Scientists think this layer started as methane either from one of Charon's cryovolcanoes or transferred from Pluto's atmosphere before being transformed into tholins by the weak sunlight.

The name of the depression is Mordor Macula, and if that sounds interesting, take a closer look at your map of the rugged surface. There are many scarps and canyons, including Serenity, TARDIS, and Nostromo Chasma. There are also Dorothy, Uhura, and Vader Craters, Kubrick Peak, and a ridge named McCaffrey. If these names seem familiar, that's because the scientists working on the New Horizons probe that flew by in 2015 decided to name features on Charon after vessels, characters, and artists linked to fictional journeys of exploration, perhaps making this destination even more enticing to visitors like you.

Address Orbiting 19,600 km (12,250 mi) from Pluto, The Kuiper Belt, The Solar System | **Getting there** Using a gravitational slingshot at Jupiter will help shorten your long trip to Charon. The downside is that you'll whizz by at high speed, so be ready to take your pictures as you fly through the system. | **Tip** Head farther out to 15760 Albion, the first trans-Neptunian object discovered after Pluto and Charon when spotted by Dave Jewitt and Jane Luu in 1992. Around 4,000 such objects are now known.

6 The Cliffs of Hathor
Close to the edge

Standing on a promontory jutting from the top of the Cliffs of Hathor, you're nervous. It's 900 meters (3,000 feet) down to the boulder-strewn floor of Hapi Valley below, the same height as El Capitan in Yosemite. You reach to check your parachute one last time and momentarily panic when you remember that you don't have one. But what good would it do anyway? There's no atmosphere here to slow you down. Help!

Relax. You're on 67P/Churyumov-Gerasimenko, the comet explored by the ESA's Rosetta spacecraft from 2014 to 2016. It's a loosely-packed ball of ice, dust, and organic molecules, and at just 4 km (2.5 mi) across, the gravity at the top of the cliffs is less than 0.001 percent of Earth's. So just follow your guide's instructions: gently fall forward and enjoy the ride. After all, it's going to take a while, about 90 minutes to reach the bottom.

That leaves you plenty of time to take in the spectacular scenery offered by the Cliffs of Hathor, named for the Egyptian goddess and mother of Horus and Ra, and also to contemplate the oddities of BASE jumping on a comet. Its center of gravity isn't directly below you, and you'll be drawn across the valley. Also, the gravity decreases by 50 percent as you descend and the comet rotates every 12.4 hours around an axis through Hapi, further complicating your timing and trajectory. All quite head-spinning.

But the valley floor is approaching now, and you need to get ready to land. It's easier than you think. After falling all that way, your touchdown speed will only be 1.3 km/h (0.8 mph), a slow amble. The next question is whether you're up for the expert challenge: leaping back to the top of the cliffs again. It involves no more effort than jumping onto a paperback book on Earth, but jump too fast and you'll escape from the comet's weak pull altogether, to drift inexorably out into the Solar System.

Address Comet 67P/Churyumov-Gerasimenko, between the orbits of Earth and Jupiter, The Solar System | **Getting there** The Rosetta probe's 10-year journey involved three flybys of Earth and one of Mars to rendezvous with 67P/C-G. You'll need to make a similar indirect approach. It is at its closest to the Sun every 6.4 years, but it's best to avoid visiting then, as the comet heats up and gets active. | **Tip** Less than 500 meters (0.3 mi) from the top of the cliffs in the Ma'at region, you'll find Rosetta's final resting place.

SOLAR SYSTEM

7 — Comet 67P/Churyumov-Gerasimenko
The adventure of a lifetime

There's no more entrancing sight in the night sky than a comet and its diaphanous tail moving through the stars. You may be lucky enough to see a few comets during your time on Earth, but imagine you could visit one – what treasures might you find? Early cometary missions were brief flybys, beginning with ESA's Giotto and Halley's Comet in 1986. The first spacecraft to orbit one was Rosetta, which rendezvoused with Comet 67P/Churyumov-Gerasimenko (67P/C-G) in 2014 and sent its Philae probe to the surface.

Discovered in 1969 by Klim Churyumov (1937–2016) and Svetlana Gerasimenko (1945–2025), 67P/C-G makes an elliptical orbit around the Sun every 6.4 years. If you visit when the comet is closest to the Sun's warmth, you'll see bright jets erupting from its heated surface. As 67P/C-G is only 4 km (2.5 mi) in diameter, its gravity is weak, and most of the material expands into a giant ball of gas and dust called a "coma." The comet casts a shadow on the inner coma, while further away, the wind and light of the Sun, combined with the comet's trajectory, separate the dust and gas into two huge tails.

Capture some of the coma, and you'll find it's about two-thirds dust and one-third water vapor and other gases, like carbon dioxide, sublimated from ice. You might be tempted to take a sniff, but beware. It'll smell of an acrid barnyard, containing trace amounts of ammonia, hydrogen sulfide, hydrogen cyanide, alcohol, and formaldehyde, along with the simplest amino acid, glycine.

Every time 67P/C-G returns close to the Sun, it loses around 0.1 percent of its mass. Within a few thousand years, either most of the ice will be gone, or the surface will be covered with a deep layer of insulating dust, preventing the Sun's warmth getting to ice beneath and making it dormant. So plan your visit sooner than later.

Address Peripatetic, from beyond Jupiter's orbit to just outside Earth's, The Solar System | **Getting there** Launched in 2004, Rosetta made a complex 10-year tour of several planets and asteroids to match 67P/C-G's eccentric orbit and rendezvous with it. | **Tip** Comet 46P/Wirtanen was Rosetta's original target before its launch was delayed more than a year due to a rocket failure and the mission sent to 67P/C-G instead. No one has visited 46P yet, so you may be the first to explore it.

8 — Dimorphos
Saving the world, one pile of rocks at a time

Around 500 million years after the Solar System formed, it's thought that the giant planets changed orbits, scattering vast numbers of asteroids onto collision courses with the planets and their moons, making many of the craters you'll see during your travels. But there's still some debris around today able to make new ones.

The last time Earth was struck by anything larger than 10 km (6.2 mi) in diameter was 66 million years ago at Chicxulub, causing the extinction of most dinosaurs. But even a 1 km (0.62 mi) object would have catastrophic consequences if it entered the atmosphere overhead or impacted anywhere near populated areas, and that happens every 500,000 years on average. How to prevent such a disaster? Head to the tiny asteroid Dimorphos and visit the site of one of our first experiments aimed at doing just that.

When you reach Dimorphos, you'll see that it's an elongated rubble pile of rock and dust about 180 meters (590 feet) long, the size of a 45-story building. Or at least it was before it was struck at over 22,000 km/h (14,000 mph) by the 580 kg (1,280 lb) NASA DART spacecraft on September 26, 2022 that took pictures until the very last second. Images taken from a smaller satellite and also from Earth showed a huge plume of material erupting at the moment of impact. Before the impact, Dimorphos took 11.9 hours to orbit 65803 Didymos, its five-times-larger asteroid companion about 1.2 km (0.75 mi) away. Since then, its orbit is 33 minutes shorter, confirming that DART had successfully moved it.

But before this technique can be applied to deflecting a small asteroid on a possible collision course with Earth, we need to know more about what Dimorphos is made of, how strongly it was held together, and how the impact changed it. So there's plenty to study while you're there. Also look for ESA's Hera spacecraft, launched in October 2024 to examine the aftermath.

Address Orbiting 65803 Didymos, Apollo Asteroid Group, The Solar System | Getting there Dimorphos and Didymos are on an elliptical orbit that takes them from near Earth's orbit to beyond that of Mars. Like Hera, you might need to make a Mars flyby to get on the right trajectory for a rendezvous. | Tip 162173 Ryugu is another member of the potentially hazardous Apollo group. Samples brought back by the Japanese Hayabusa2 spacecraft show that it may be an extinct comet.

SOLAR SYSTEM

9 Dust Devils of Mars
Wandering graffiti artists

When you visit the desert of Mars, you won't find it a particularly windy place on average, thanks to the wan sunlight, lack of oceans, and thin, dry atmosphere. The wind can occasionally pick up, reaching speeds of 100 km/h (62 mph). Over time, it can blow sand into ripples, dunes, barchans, and streaks behind obstacles like craters. But because the surface pressure on Mars is only one percent of that on Earth, you'll feel a gentle breeze on your spacesuit at best, not the kind of violent blow that ended up stranding Mark Watney in the 2015 film *The Martian*.

He and his crewmates encountered one of Mars' periodic dust storms, when a wall of sand sweeps across the region, similar to a haboob. Every few years, these storms become large enough to envelop the whole planet for weeks, endangering robotic rovers and landers. Again, the wind isn't the main problem, but rather the fine, electrostatically-charged dust particles that get everywhere, sticking to solar panels and reducing power, penetrating gears, and potentially damaging other vital equipment.

Ironically, another kind of martian wind can come to the rescue. During the morning, as the heat of the day begins to build up and drive convection, you may see a column of dust wandering across the desert. These are dust devils, and they can reach heights of anywhere between 0.5 to 2 km (0.3–1.25 mi), taller than most tornadoes on Earth but much less dangerous. If one passes over your dust-covered vehicle, it can provide a free cleaning service. In some places, dust devils are very common, scouring away the surface dust and creating beautiful tangles of dark trails behind them. You can use the size and shapes of these dust devil trails to help you figure out the prevailing wind direction, among other things. If you see one coming towards you, you probably won't be able to outrun it – just brace yourself and enjoy the moment.

Address Hooke Crater on Mars (44.6°S, 315.2°E), Fourth Planet from the Sun, The Solar System | **Getting there** Mars makes its closest approach to Earth every two years at a distance of around 60 million km (38 million mi), so time your trip accordingly. | **Tip** Hooke lies on the northern edge of the 1,700 km (1,100 mi) wide Argyre basin, created by a 200 km (124 mi) impactor around four billion years ago. It may have been filled by a giant lake for millions of years after (49.7°S, 316.0°E).

SOLAR SYSTEM

10   Europa
An icy world in motion

As you plan your journey, you'll learn that Europa, the smallest of Jupiter's four Galilean moons, is the smoothest body in the Solar System. However, as you approach, you'll be amazed by the enormous variety of features on its surface: brown ridges and cracks, rounded spots, puddles and lobes, small hills and hummocks, and chaotic terrain, all on a blue-white background. You'll also notice that Europa has very few craters. That doesn't mean that it hasn't been hit by asteroids like every other Solar System body. But Europa is covered in a crust of ice floating atop a water ocean twice the volume of all of Earth's oceans combined, allowing the surface to be constantly renewed, erasing any craters.

The heat needed to keep the water below the icy crust in a liquid state comes from tides created in Europa's ocean. These tides warm the icy crust and the rocky ocean floor far below via friction. You might think the main cause is Jupiter, 670,000 km (420,000 mi) away, but two of the other moons, Io and Ganymede, are probably more to blame. Orbital resonances between them create tides in each other's oceans, lifting, cracking, and refreshing the warmed crust. The crust can also fracture into great icebergs that move around in the sea of ice.

Hubble Space Telescope observations show water vapor emerging from surface cracks, so the ice might be only a few kilometers thick in places, while the water below could be anything from 40 to 150 km (25–94 mi) deep. NASA's Europa Clipper is en route to learn more about the moon and study its ocean from orbit. One big question is whether the conditions exist for life to have evolved in the ocean. But to look into this and many other mysteries, Clipper will have to survive the intense radiation that Europa experiences close to Jupiter, as will your spacecraft. It could kill you in just a few days, so plan for a short stay.

Address In orbit around Jupiter, Fifth Planet from the Sun, The Solar System | **Getting there** Jupiter is 780 million km (490 million mi) from the Sun. Similar to Europa Clipper, you can plan a route involving flybys at Mars and Earth to minimize fuel use on your way there. | **Tip** Pasiphae is the largest of a group of almost 20 moons making elliptical orbits far from Jupiter and in the opposite direction of the planet's rotation. They're remnants of a captured, fractured asteroid.

SOLAR SYSTEM

11 The "Face" on Mars
Much ado about nothing

In 1976, the NASA Viking 1 orbiter flew high above Cydonia on Mars. Located between the northern plains and southern highlands, the region contains many valleys, hills, and mesas. In a handful of fuzzy pictures, one of those mesas was side-lit by the Sun and appeared to show human features. This face on Mars became a cause célèbre among credulous space fans who claimed that it was just one of many artificial features in a city built by a long-lost civilization. Despite being dismissed by Viking mission scientists as a trick of the light and an example of pareidolia that also leads us to see a human face in the Moon, for example, the speculation continued, fueling many articles, books, TV shows, and films.

Fortunately, you can now follow in the footsteps of later space missions and see that the scientists were right all along. Looking down from martian orbit and armed with a better camera, you'll immediately see that the mesa is a perfectly natural geological feature. It is roughly 2.6 by 1.9 km (1.6 by 1.2 mi) in size and 250 meters (820 feet) high, crossed by fissures and landslides created as the mesa eroded and slumped into the surrounding terrain. It's thought that Cydonia was once home to ancient seas or lakes, later covered by lava and sediment. Subsequent erosion by water formed the many mesas and debris-filled valleys.

Descend to the surface of Mars to climb the mesa and consider its history, both real and imagined. The Universe has many amazing sights to see and fascinating stories to tell, and human curiosity and ingenuity are revealing many of its secrets. But as Carl Sagan (1934–1996) warned us in his 1995 book, *The Demon-Haunted World*, we should be skeptical of wishful thinking and also of disingenuous people happy to sell us made-up stories and conspiracy theories about advanced extraterrestrials, whether on Mars or in the skies above us.

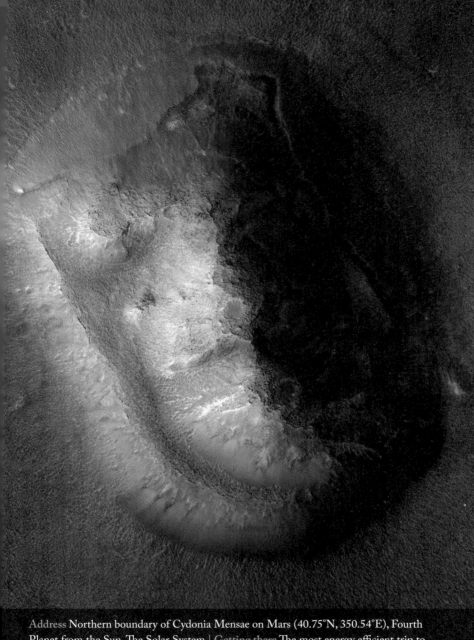

Address Northern boundary of Cydonia Mensae on Mars (40.75°N, 350.54°E), Fourth Planet from the Sun, The Solar System | **Getting there** The most energy efficient trip to Mars would use Hohmann transfers – 9 months outbound, 16 months at the planet until Mars and Earth aligned again, then 9 months home. | **Tip** To the west, you'll find Cydonia Labyrinthus, a flatter region covered with many intersecting valleys that carve the terrain into irregular polygons (41.3°N, 347.4°E).

SOLAR SYSTEM

12   Far Side of the Moon
Not your father's prog rock album

If you're searching for some peace and quiet away from our home planet, then look no further than the far side of the Moon. It's less than 400,000 km (250,000 mi) away and yet completely hidden from noisy Earth.

We only see one side of our nearest companion. Soon after it formed, the Moon became "tidally locked." As it makes its month-long orbit around Earth, it also spins on its own axis at the same time. That means one side is permanently facing us, apart from slight wobbles, called "librations," while the other is concealed. But Pink Floyd notwithstanding, it's not dark on the far side, at least no more so than on the near side. Both sides experience a fortnight of scorching daylight up to 120°C (248°F) followed by a frigid two weeks of night down to -171°C (-276°F), with no atmosphere to smooth out the fluctuations.

The far side is more heavily cratered and has few of the large, dark maria, or "seas," found on the near side. When the Moon formed, it was at just 5–10 percent of its current distance to Earth, and this proximity to our hot, young planet affected the composition of the Moon's near side. Earth's heat led to a thinner crust that was more easily punctured by asteroid impacts, letting hot lava flood out. The far side ended up with a thicker crust and was spared that fate.

The only humans to have seen the lunar far side are 24 of the Apollo astronauts and then only from above, as all landings were on the near side. But Chinese spacecraft have landed and returned samples from the far side recently, and there's discussion about building a radio observatory there. Shielded from the blare of Earth's incessant technological chatter, the faintest cosmic whispers would become audible. Thanks to the same tidal forces that locked its two faces, the Moon is receding from Earth at 3.8 cm (1.5 in) per year, so every day you delay, the longer the journey will get.

Address Orbiting Earth, Third Planet from the Sun, The Solar System | **Getting there** The Moon is our closest cosmic neighbor, just three days away, and was first visited by humans from 1968–1972. It's envisioned that regular transport there will start again soon. | **Tip** A less dusty location for your radio telescope would be the Earth-Moon L2 point, 62,800 km (39,000 mi) above the far side. Shadowed by the Moon, it too is free of Earth's racket.

13 Ganymede
Some moons are more equal than others

Look up at Jupiter with binoculars or a small telescope, and you'll likely see faint points of light in a straight line on either side of the planet. The points will move from hour to hour, night to night as they circle the planet.

This was the sight that greeted Galileo Galilei (1564–1642) and Simon Marius (1573–1625) when they turned their newly invented telescopes to the heavens in 1609 and 1610, respectively. In his book *Sidereus Nuncius*, or "starry messenger," Galileo named the four points the Medician stars to honor his patron Cosimo II de' Medici and his three brothers. We now know them as Io, Europa, Ganymede, and Callisto, or the Galilean moons. Their discovery proved that Earth is not the center of the Universe around which all objects orbit, and Galileo's subsequent promotion of the heliocentric model caused him much grief with the Catholic Church.

Visit any of those moons today, and you'll see things Galileo could only have imagined. Take Ganymede, for example, orbiting a million km (620,000 mi) from Jupiter. It's the largest moon in the Solar System, even larger than Mercury. You'll see that Ganymede's surface is broken into strange, angular regions. Some are darker and more cratered, suggesting that they're older. Lighter parts with fewer craters and covered in grooves are icy regions kept fresh by tidal heating and flexing at some point in the past.

Ganymede is the only moon with a magnetic field, and scientists have used observations of its aurorae to confirm a global saltwater ocean 50–200 km (31–125 mi) below its icy crust. The ocean may descend 800 km (500 mi) to a floor of high-pressure ice. Farther below, it's believed there's a rocky mantle that warms the liquid ocean through radioactive decay, and deeper yet, an iron-nickel core. ESA's JUICE mission is en route to Jupiter and will orbit Ganymede in the mid-2030s to study this ocean world from above.

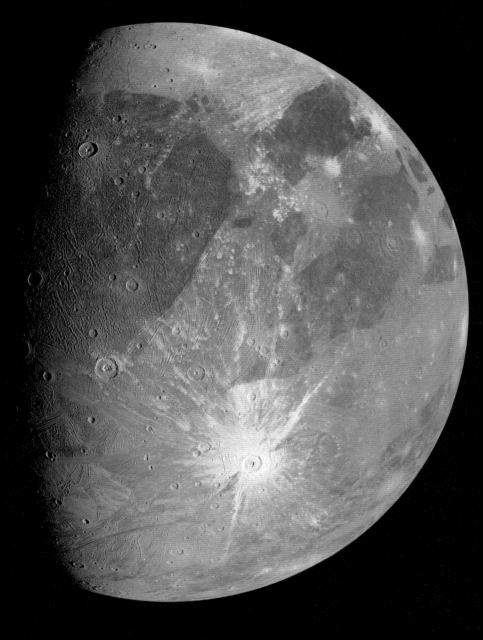

Address Orbiting Jupiter, Fifth Planet from the Sun, The Solar System | **Getting there** You'll need to plan a complex trajectory to get to Jupiter and later into orbit around Ganymede. JUICE is using flybys at Earth, the Moon, and Venus, for example. | **Tip** Metis, Adrastea, Amalthea, and Thebe are small moons that help maintain Jupiter's faint ring system. But they're deep inside Jupiter's lethal radiation belts, so it'd be a very dangerous side trip.

SOLAR SYSTEM

14 Hubble Space Telescope
Beauty is eternity gazing at itself in a mirror

To explore the work of a portrait photography pioneer, you might visit Dimbola, the studio of Julia Margaret Cameron (1815–1879) on the Isle of Wight in the UK. Similarly, to witness the place where many iconic astronomical images have been taken, you should plan a trip to see the Hubble Space Telescope.

The first ideas for a telescope in space, above the turbulence, glow, and absorption of Earth's atmosphere, came from rocket engineer Hermann Oberth (1894–1989) in the 1920s and astronomer Lyman Spitzer (1914–1997) two decades later. Spitzer and Nancy Grace Roman (1925–2018) tirelessly promoted a large space telescope project, and development started in the late 1970s.

This became the NASA/ESA Hubble Space Telescope, which launched in 1990 on the space shuttle Discovery. It was named for Edwin Hubble (1889–1953), whose observational evidence for an expanding Universe set the scene for much of the telescope's studies of the cosmos. It was not all plain sailing initially, as Hubble's main mirror had been mistakenly polished into the wrong shape, but astronauts installed corrective optics in 1993 and refurbished other parts during four later space shuttle missions. Over more than three decades, the telescope's discoveries have impacted all of astronomy, from the Solar System to exoplanets, star birth, and distant galaxies.

However, the window for your visit may be closing. Hubble is aging, its orbit is slowly decaying. Left to its own devices, it will fall back to Earth sometime in the next decade. To ensure that it won't hit any populated areas, a rendezvous mission is planned to direct it safely to Point Nemo in the South Pacific. Another option is to lift it to a higher orbit, where it would survive for thousands of years. If you do visit while Hubble still is in space, stay well back, as it will likely be taking more glorious pictures and spectra of our Universe.

Address Low Earth Orbit, Third Planet from the Sun, The Solar System | **Getting there** Hubble orbits at around 500 km (311 mi) above the Earth, roughly the same distance between New York City and Buffalo, or Cologne and London. Your rocket can reach it in just a few hours. | **Tip** The NASA Fermi Gamma-ray Space Telescope studies black holes, neutron stars, and pulsars. It's in a similar orbit to Hubble, so if you time things right, you may be able to spot it passing only a few kilometers away.

15 Hyperion
A spongy, cosmic beehive

During your travels, you'll become very familiar with the craters of different sizes that cover many of the rocky and icy moons and planets. They are the signatures of impacts by material left over from the formation of the Solar System billions of years ago. But there's nowhere quite as strange as Hyperion, a small world so cratered that it has taken on a weird, sponge-like appearance.

Discovered in 1848, Hyperion is a surprisingly irregular moon of Saturn with an average diameter of 270 km (168 mi). At a temperature of about -180°C (-292°F), it's made of a mixture of ices, mostly water with a spritz of frozen carbon monoxide, plus a dash of rock. Repeated flybys of the NASA/ESA/ASI Cassini spacecraft revealed that its interior is porous, perhaps 40 percent empty space, giving it a low density, about half that of ice. That leads to unusual behavior when it's hit by an impactor. A crater forms by compressing the surface rather than gouging material out, and thanks to a surface gravity that is just 0.2 percent that of Earth's, the little that is ejected mostly escapes the moon altogether and heads into orbit around Saturn.

Over time, Hyperion has become completely pockmarked with an irregular honeycomb of these odd craters. The largest is 120 km (75 mi) across and 10 km (6 mi) deep, and many are deep and sharp-edged, with evidence of landslides. Some have a mysterious dark coating on their chilly floors. It's thought this might be a mixture of carbon and hydrogen-bearing molecules that perhaps came originally from Phoebe or Titan, other moons of Saturn.

Beneath the surface, Hyperion's interior might be uniform and fluffy, but it could contain vast caverns. If you're tempted to explore them, you'll need to be a skilled pilot to land on Hyperion. It's locked into a resonance with nearby Titan, and its spin changes chaotically as it orbits Saturn.

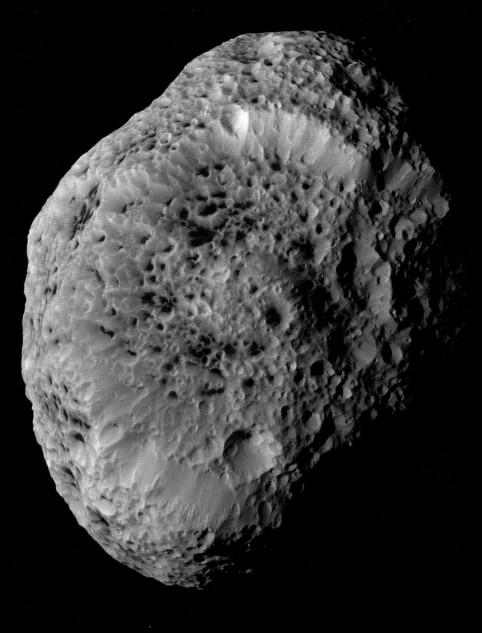

Address Orbiting Saturn, Sixth Planet from the Sun, The Solar System | **Getting there** Flybys using Venus, Earth, and Jupiter can provide an efficient route to reach Saturn, 1.4 billion km (870 million mi) from the Sun. | **Tip** By contrast, Rhea, Saturn's second-largest moon, is boringly round, but has a bright, fresh, water-ice crater called Inktomi you ought to visit.

16 International Space Station

An accessible, if expensive, tourist destination

Many locations in space will probably never be reached by humans, but there is one to which you could book a trip tomorrow. The International Space Station circles Earth 16 times a day at 27,600 km/h (17,100 mph). At that speed, our spherical planet curves away from the ISS at the same rate the ISS falls towards it under gravity, resulting in the ISS orbiting at an altitude of around 400 km (250 mi).

The first Russian and American components of the ISS were delivered in 1998 and since then it has expanded to the size of a football field, with habitation and propulsion modules, giant trusses and solar arrays, and usually several docked crew and cargo vehicles. It has been continuously occupied since 2000. Hundreds of astronauts and cosmonauts from more than 20 countries have visited the outpost, many of them several times, some spending up to a year onboard. While most work for national and international space agencies, an increasing number of private individuals are now also making the trip. To do so, you'll need deep pockets or a wealthy friend. For now at least, expect to pay over $50 million for transport and accommodation.

When you get to the ISS, you'll find that it's a science lab, loaded with experiments probing the properties and behaviors of materials from mineral to biological under microgravity conditions. We're also studying the Universe from this special vantage point. A key goal is to understand the effects of long-term space travel on humans, essential to plans to establish a permanent base on the Moon and travel to Mars.

The ISS will likely be deorbited into Point Nemo in the Pacific Ocean in the coming decade. What will replace it? Some are planning private space stations and even hotels. But whether travel to them will become truly affordable remains to be seen.

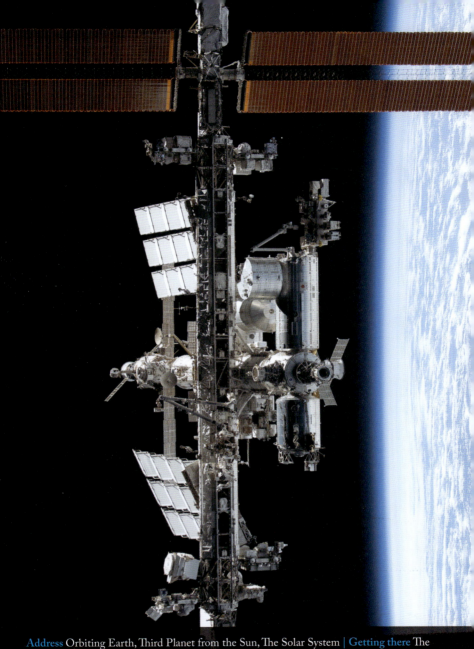

Address Orbiting Earth, Third Planet from the Sun, The Solar System | **Getting there** The ISS is currently reachable via three different rockets and crew vehicles. The quickest you can expect to rendezvous and dock is around three hours after launch, although it can take up to a day. | **Tip** Construction of the Chinese space station, Tiangong or "Heavenly Palace," started in 2021 and it may also become available as a destination for space tourism in the future.

17 Io
A land of ice and fire

When you travel to Jupiter, you'll find yourself more than five times further from the Sun than when you're on Earth, and sunlight is almost 30 times weaker. So it's cold out there, with the surfaces of the icy moons Europa, Ganymede, and Callisto at least 100°C (180°F) below the freezing point of water even during the daytime. It's even more surprising, then, that some parts of the surface of Io, the fourth large Galilean moon, reach a scorching 1,300°C (2,370°F). The reason? Volcanoes.

Io is a little smaller than Earth's Moon, but even from afar, you'll realize that something strange is going on there. The rugged surface has broad plains covered in frost, and isolated mountains taller than Everest. It's a colorful, blotchy mix of whites, reds, and yellows. Black patches mark the calderas of volcanoes, occasionally erupting and funneling hot basalt from its rocky interior. Vast sulfurous plumes rise hundreds of kilometers above the volcanoes and led to their surprise discovery in 1979, when Linda Morabito examined images of Io taken during the Voyager 1 flyby.

If you tune your goggles to the infrared, you'll see hot spots all over Io, and it's thought that there may be 400 active volcanoes on the moon, many now named for deities associated with fire, including Loki, Pele, and Prometheus. Lava flows and deposits from the plumes constantly renew the surface of Io and remove any signs of craters. If you do decide to risk a visit down there, you'll need to tread very carefully indeed.

So far from the Sun, what's the source of all this heat? Io is the closest of the four large moons to Jupiter and is pulled by the gravity of the planet, Europa, and Ganymede. Tidal forces squash and stretch the moon's interior, heating it up and creating a sub-surface lava ocean that powers Io's volcanoes, the most geologically active place you can visit in the Solar System.

Address Orbiting Jupiter, Fifth Planet from the Sun, The Solar System | **Getting there** Journeys to the outer planets can be complex. The JUICE mission will use three flybys of Earth, one of the Moon, and one of Venus during its eight-year cruise to reach Jupiter. | **Tip** Some material ejected by the volcanoes escapes Io to form a dense plasma torus around Jupiter. Intense magnetic fields channel some of that to its poles, where it creates a bright spot in the planet's aurora.

SOLAR SYSTEM

18 Jupiter
King of the planets

Head to Jupiter, and you'll earn bragging rights for the biggest vacation ever. With a mean diameter of 140,000 km (90,000 mi) and more than double the mass of all of the other planets in the Solar System combined, it's a world of superlatives and rightly named for the chief god in the Roman pantheon.

At 780 million km (490 million mi) from the Sun, Jupiter lies beyond the main asteroid belt and is unlike any of the four planets inside it. Where the other planets have rocky surfaces and varying amounts of atmosphere, Jupiter is all atmosphere and no surface. It's mostly made of hydrogen and helium, with traces of carbon, oxygen, sulfur, water, and ammonia. At the cloud tops, the pressure is the same as on Earth, but the temperature is a chilly -110°C (-166°F). Should you be brave enough to attempt a descent through the troposphere, the pressure and temperature will rise quickly. Sooner or later, weird things will happen, as hydrogen turns smoothly from gas into a supercritical fluid and, deeper still, into a liquid metal. Jupiter doesn't even have a simple rocky core – its heavy elements have partly dissolved in the metallic hydrogen and diffused outwards, perhaps thanks to an impact when the planet was forming billions of years ago.

Back above the atmosphere, you'll see that Jupiter is rotating rapidly, and each day is less than 10 hours long. But because it's not solid, its rotation speed differs with latitude, leading to its many colorful belts, whorls, and storms. The most famous of these, the Great Red Spot, has lasted for centuries, its distinctive color likely from chemical reactions between ammonia and acetylene. But it's shrinking and may disappear within decades, while other storms come and go. If you can drag your eyes away from the ever-changing clouds, there's much more to see here: aurorae at both poles, a faint ring system, and almost 100 moons.

Address Fifth Planet from the Sun, The Solar System | **Getting there** Arriving in orbit around Jupiter can be tricky and time-consuming. Oddly, it may be helpful to head towards the Sun first to make a flyby at Venus, changing onto a better orbit. | **Tip** Jupiter has an intense magnetic field that creates a giant cavity in the solar wind. The tail of this "magnetosphere" extends up to 500 million km (310 million mi), making it one of the largest structures in the Solar System.

SOLAR SYSTEM

19 Lutetia
Heavy metal in the sky

Meteorites are space rocks of many different shapes, sizes, and compositions that make it through Earth's atmosphere without completely burning up. Most common are the stony meteorites known as "chrondrites", made up of clumps that accreted out of dust and small grains in the early Solar System. Occasionally you'll find rocks from the Moon or even Mars, blown off their surfaces by giant impacts. Then there are meteorites mostly made of metal, typically an iron-nickel alloy, which account for less than 6 percent of all known meteorites. Where did they come from? If you head to asteroid 21 Lutetia, you might find some answers.

As you approach, you'll see that Lutetia is irregular, cratered, and fairly large, with an average diameter around 100 km (63 mi). From a distance, the surface looks soft, thanks to a layer of dust 3 km (2 mi) thick. Like other asteroids in its class, Lutetia has a reddish-brown color and an overall spectrum similar to that of iron meteorites. That connection leads astronomers to believe that asteroids like this one have a high metal content, giving them the name M-type. It's thought that all but the largest M-types are fragments of the dense, metal-rich cores of protoplanets that were shattered by giant collisions when our Solar System was young.

While you're digging for more clues inside the asteroid, consider amateur astronomer Hermann Goldschmidt (1802–1866), who discovered Lutetia in 1852 through the window of his garret above a famous brasserie less than 1 km (0.6 mi) from Notre Dame Cathedral in Paris. The name Lutetia derives from the Roman name for the city. Goldschmidt used a small telescope bought with the proceeds of selling some portraits of Galileo and later won prizes and plaudits as the most successful discoverer at the time of small planets, as asteroids were then called. He found 14. It's fitting that he now has one named after him: 1614 Goldschmidt.

Address The Asteroid Belt, between Mars and Jupiter, The Solar System | **Getting there** You'll be retreading the path taken by the Rosetta probe which flew past in 2010. This involved three flybys at Earth and one at Mars before entering the asteroid belt. | **Tip** 16 Psyche is the largest M-type asteroid, more than double the size of Lutetia. The NASA mission of the same name will be visiting and going into orbit around it in 2029.

SOLAR SYSTEM

20 Maat Mons
Is Venus still geologically active?

Orbiting above Venus' dense, sulfuric acid clouds, there is nothing to be seen of the surface far below. Set your goggles to radar mode, and a complex terrain is revealed. There are gently-rolling uplands with occasional highlands and mountains sprinkled with metallic snow, and flat basalt-flooded lowlands, interrupted by many faults, folds, fissures, and wrinkles. With surface temperatures approaching 465°C (870°F), it's no surprise that Venus lacks oceans, but there's also no sign of the plate tectonics that make mountains, trenches, and earthquakes on Earth. What Venus does have, though, is volcanoes. Around 85,000 of them.

The surface of Venus has been shaped by dramatic events. There are many craters resulting from impacts, but none smaller than 2 km (1.25 mi) in diameter, probably because smaller objects are pulverized by the planet's incredibly dense atmosphere. And none of the larger craters are more than 500 million years old, suggesting that Venus was resurfaced then in a planet-wide paroxysm of volcanic activity. That's relatively recent in cosmic terms and begs the question of whether Venus might still be geologically active today.

Maat Mons, named for Ma'at, the Egyptian goddess of truth and justice, is a huge shield volcano rising 5 km (3.1 mi) above the surrounding plains. In the 1990s, the NASA Magellan mission made radar images showing old lava flows around its large caldera. However, revisiting the data more than 30 years later, scientists noted that a large vent on the volcano's flank had doubled in size in just eight months in 1991. They also spotted a fresh lava flow nearby and concluded that lava had filled the vent before spilling over and running downhill. This is direct evidence for ongoing volcanism on Venus. So, if you feel bold enough to make a visit to Maat Mons, be sure to keep your wits about you – things may suddenly get even hotter than you bargained for.

Address Atla Regio on Venus (0.5°N, 194.6°E), Second Planet from the Sun, The Solar System | Getting there Venus comes closer to Earth than any other planet, so if you time it right, you'll only have 38 million km (24 million mi) to cross. Get it wrong, though, and you're in for a much longer journey. | Tip Sif Mons is another venusian volcano that's believed to be still active, although it's a bit of a trek, because it's on the other side of the planet (22.0°N, 352.4°E).

SOLAR SYSTEM

21 Mars
The planet of dreams or nightmares?

You've seen the photos and movies, and heard the siren call of the glossy tourist brochures. When you reach Mars, they tell you, it'll be like a trip to one of Earth's beautiful deserts: stark and rugged, sand dunes and buttes, canyons and mountains, blue skies overhead. The Red Planet holds an almost mythical attraction for would-be extraterrestrial adventurers and homesteaders. Except the truth is that Mars isn't a nice place at all for humans, and it will try to kill you.

Mars is a small world, with just 38 percent of Earth's gravity, and, at around 78 million km (48 million mi) further from the Sun than our planet, also a cold one. Yes, it can reach 20°C (70°F) on a summer's day at the equator, but it drops to -75°C (-103°F) at night and is 50°C (90°F) colder at the poles. One reason for these extremes is the very thin atmosphere, only one percent of the density of ours and mostly carbon dioxide. There's precious little oxygen to breathe, and the pressure is so low, you'd need a spacesuit to avoid the air in your lungs and dissolved gases in your blood expanding rapidly. The thin atmosphere and lack of a global magnetic field let dangerous radiation reach the surface. Breathing the pervasive, rusty, red dust would shred your lungs. As for water, Mars has plenty of ice at the poles and perhaps buried glaciers at lower latitudes, but unlike Earth's deserts, you'll find no fertile oases here.

It's still an exciting destination to explore though, as long as you're careful. You'll see amazing things from orbit and on the surface. Mars also holds a vital secret for scientists. It was a very different world when younger – with a thicker, warmer atmosphere and a surface covered with lakes and oceans. Perhaps primitive microbes evolved and got a toehold – they might have left fossils or may even still be living underground. Searching for evidence of life is a worthy goal for robot rovers and well-equipped space adventurers like you.

Address Fourth Planet from the Sun, The Solar System | **Getting there** Mars takes 687 Earth days to orbit the Sun. If you calculate it right, that means Mars and Earth come closest every 26 months, making that the interval between the best opportunities to start your roughly nine-month-long journey to the Red Planet. | **Tip** Nearly 20 spacecraft have orbited Mars since 1971, and many are still up there. It might be interesting to seek some out, but please don't go near any currently doing science.

SOLAR SYSTEM

22 Mercury
The Solar System's problem child

A world of extremes and contradictions, Mercury is an excellent travel destination. But it's surprisingly hard to get to and challenging once you arrive. In both cases, the reason is simple – Mercury is the closest planet to the Sun. For example, if you just head straight there and let the Sun's gravity pull you inwards, by the time you reach Mercury, you'd be moving too fast to get into orbit. Instead, you'll need to take a circuitous route involving flybys at Earth, Venus, and Mercury itself. These encounters will slow you down, allowing you to reach your destination at just the right speed.

Once there, you'll be exposed to sunlight that's 10 times brighter than on Earth. During daytime, the surface of Mercury can reach 450°C (842°F), as hot as a pizza oven, but without an atmosphere, it plummets to -180°C (-292°F) at night. Conversely, there are craters at the north and south poles that never see sunlight and have ice in them, perhaps detritus from comets that crashed there.

Most of Mercury is dark gray, thanks to volcanic eruptions that flooded the surface with lava billions of years ago. The planet has been shrinking ever since, wrinkling like a dried orange. Curiously, Mercury has a magnetic field, about one percent as strong as Earth's. This normally requires a planet's metal core to be liquid, but scientists believed Mercury's small core should have frozen long ago. The answer may be linked to the fact that the planet is also very dense. One theory is that it suffered a giant impact, stripping away the outer layers and leaving a mostly iron core that's still molten. Or perhaps it's just that Mercury formed in a weird way, adding to its mystery.

Expect your body clock to get quite confused during your visit. A year is how long it takes a planet to orbit the Sun, and a day is the time from one midnight to the next. On Earth, one year lasts 365.25 Earth days. But Mercury orbits the Sun quickly and spins slowly, with the bizarre result that one Mercury day is two Mercury years long.

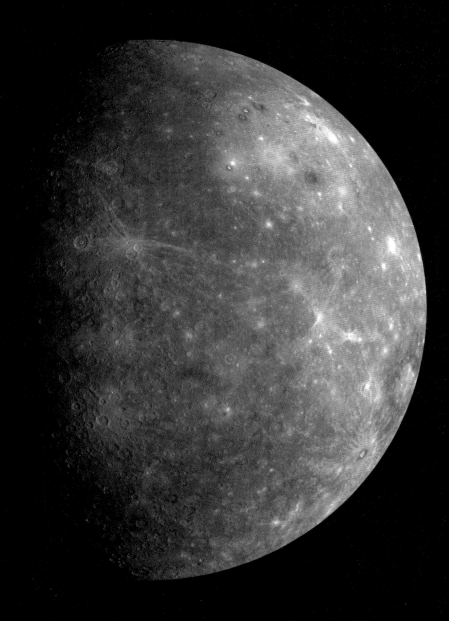

Address First Planet from the Sun, The Solar System | Getting there On average, Mercury (no, not Venus or Mars) is the closest planet to Earth. But the roundabout route needed to go into orbit will take a long time. It's taking eight years for ESA's BepiColombo spacecraft to make this journey. | Tip Once you're at Mercury, be sure to look back towards Earth. You may discover some new, inner Solar System asteroids that are impossible to see from Earth because of the glare from the Sun.

SOLAR SYSTEM

23 Miranda
A paradise for space geologists and daredevils

Miranda is one of the most fascinating moons of Uranus, all of which are named for characters in works by Pope or Shakespeare, in this case the latter's *The Tempest*. Discovered in 1948 by Gerard Kuiper, Miranda is just 470 km (290 mi) in diameter, and is the smallest moon that's roughly spherical in the whole Solar System. It's the smallest moon that's roughly spherical in the whole Solar System. That means it has enough mass to keep it flexible internally and allow gravity to pull it into a ball, while smaller moons tend to be irregular. But once you get here, you'll find that it's all the departures from spherical that make Miranda so interesting.

From above, you'll see rugged and fractured terrain, a patchwork quilt delineated and crossed by faults, gorges, ridges, and craters. Some regions look as though they've had a giant garden rake pulled through them. One possibility is that Miranda suffered one or more huge impacts in its early history, causing it to break up and then badly reassemble under its own gravity, like Frankenstein's monster. Or it may have been kneaded, heated, and reshaped over billions of years by tidal forces, thanks to Uranus and some of its other moons. It's even possible that a sub-surface water ocean was involved and is still liquid.

Descend to the surface, and you'll discover perhaps the most remarkable feature on Miranda: the giant cliffs of Verona Rupes. Although estimates of the height of this escarpment vary wildly from 5 to as much as 20 km (3.1–12.5 mi), they're probably the highest sheer cliffs in the Solar System and are sure to attract gawkers. Future space BASE jumpers might also want to visit after conquering the Cliffs of Hathor on Comet 67P/Churyumov-Gerasimenko. Falling from 10 km (6.2 mi) up, you'd take more than eight minutes to reach the surface. But Miranda is large enough that you'd land at about 140 km/h (88 mph), so you'll need to activate a giant airbag to cushion your landing.

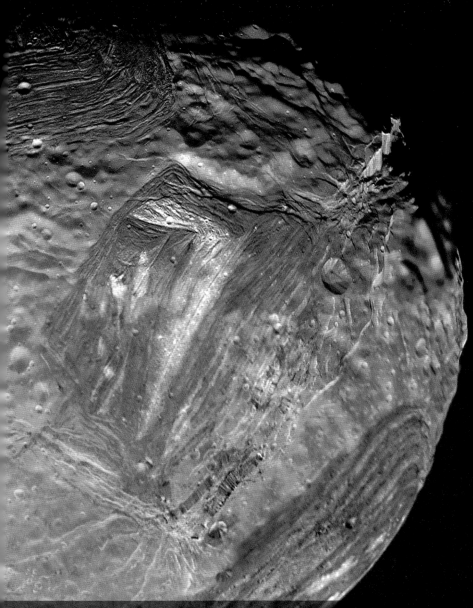

Orbiting Uranus, Seventh Planet from the Sun, The Solar System | Getting there
Uranus is 2.9 billion km (1.8 billion mi) from the Sun on average. Gravitational slingshots at Jupiter and Saturn can help you on your way if you launch when the planets are optimally aligned. The rings of Uranus lie inside the orbit of Miranda, making for an easy side trip. Most of the 13 rings are narrow, thin, and very dark, likely a mix of ice and organic compounds.

24 — Near Side of the Moon
A cosmic treasure island?

From Earth, we can only see one side of the Moon. A patchwork of grays, the lighter cratered highlands are juxtaposed against the lava-flooded darker maria or "seas." These features have been mythologized across millennia of human history. Some see a human face in the full Moon. East Asian and Mesoamerican cultures recognize a rabbit. The Haida people of the Pacific Northwest draw a boy carrying sticks. And in the southern hemisphere, Māori legends talk of a woman and a tree etched on the lunar surface.

But even the most familiar destinations can hold surprises when viewed in a new light. If you make the short journey to our nearest neighbor and examine the dusty lunar regolith in detail, you'll see subtle colors signaling the presence of different minerals. For example, the bluer a region is, the higher the fraction of titanium, while redder regions contain more iron. Mixed with these minerals is something more exotic: helium-3. This isotope has been carried by the solar wind and trapped in the lunar surface over billions of years. Some envision mining and using it for nuclear fusion reactors on Earth in the future.

A more basic but critical resource for long-duration visitors is water for drinking, but also to break into oxygen for breathing and rocket fuel. The Moon was long believed to be bone dry, but we now know that there are permanently-shadowed craters near the poles which remain cold enough for water ice to exist. There's even water in sunlit regions – about a soft drink can's worth per cubic meter of regolith made as hydrogen from the Sun combines with oxygen in the Moon's surface with the help of micrometeorite impacts.

The presence of these resources is one reason for a resurgence in interest in the Moon. When you arrive here, the question is whether it will be the "magnificent desolation" of Apollo 11's Buzz Aldrin or a gigantic, despoiled strip mine.

Address Orbiting Earth, Third Planet from the Sun, The Solar System | **Getting there** At a distance of around 384,400 km (238,855 mi), the Moon is a relatively short three-day journey from Earth. Getting above Earth's atmosphere and gravity is the hardest part. | **Tip** Every few years, small asteroids from the Arjuna family become captured by Earth's gravity for a week or so. Even though they're only tens of meters in size, it might be fun to visit another of Earth's temporary moons.

25 Neptune
Still the most distant planet in the Solar System

Although our Solar System extends far beyond the orbit of Neptune, it's the last giant planet that you'll encounter on any journey towards interstellar space – probably.

Tracking Uranus across the sky, astronomers realized that it wasn't quite following its predicted orbit and proposed that perhaps another planet farther from the Sun was tugging on it slightly. Mathematicians Urbain Le Verrier (1811–1877) and John Couch Adams (1819–1892) made independent calculations of where this body might lie, and in 1846, Johann Gottfried Galle (1812–1910) spotted a new, slowly moving object close to Le Verrier's coordinates. It was later found that Galileo Galilei (1564–1642), among others, had likely observed it much earlier without recognizing its significance. After a rather unseemly tussle among astronomers and nations, the new planet was named Neptune for the Roman god of the sea.

As you approach, you'll see that it is encircled by faint, fragmented rings and 16 moons. You'll note the pale, blue-green shade of its hydrogen-helium atmosphere, colored by hints of methane like on Uranus. But Neptune is 50 percent further from the Sun, and its upper atmosphere is one of the coldest places in the Solar System, dropping to a decidedly chilly -218°C (-360°F). There are giant storms, and frozen methane clouds scud along in winds reaching 2,200 km/h (1,400 mph). Diving deeper, you'll feel the temperature and pressure rising inexorably. About 7,000 km (4,350 mi) down, you might encounter diamond hailstones. Closer to the core are bizarre forms of highly conductive, solid water ice at temperatures similar to the surface of the Sun.

Some astronomers believe that apparent anomalies in the locations of small, icy bodies beyond Neptune might be due to a vastly more distant Planet Nine. If this planet truly exists and you're lucky enough to encounter it on your trip, take pictures.

Address Eighth Planet from the Sun, The Solar System | **Getting there** Neptune is 4.5 billion km (2.8 billion mi) from the Sun on average. Gravitational slingshots at Jupiter, Saturn, and Uranus can help you get there more quickly if you launch when the planets are best aligned. | **Tip** Visit Hippocamp, one of the smallest of Neptune's known moons at just 35 km (22 mi) in diameter. It wasn't seen by NASA's Voyager 2 during its 1989 flyby, but it was spotted in Hubble images in 2013.

SOLAR SYSTEM

26 North Polar Hexagon
Who's a pretty polygon then?

Visitors to Saturn often focus on its equatorial regions to view the planet's magnificent ring system. Beneath the rings at these latitudes, Saturn's upper atmosphere is a set of belts and zones in a bland, beige palette, and the uppermost clouds made of ammonia ice at around -250°C (-418°F) are battered by ferocious winds reaching 1,800 km/h (1,100 mph). Warmer cloud decks of ammonium disulfide and water ice lie hundreds of kilometers deeper.

But the real excitement takes place in Saturn's polar regions, particularly in the north. As your eye moves to higher latitudes, you'll see small storms drifting polewards, and there will be clearer stripes and dark spots. And then suddenly, at around 78°N, you'll notice something quite bizarre surrounding the pole: an enormous hexagon spanning 29,000 km (18,000 mi), each of the six sides larger than Earth. Inside the hexagon are many storm systems, large and small, and as you move closer to the pole itself, ragged clouds hurtle around a giant vortex at speeds of up to 600 km/h (375 mph) at the edges. Imagine yourself in the calm center of this cyclone, 9,000 km (5,600 mi) wide, and looking at its eyewall descending more than 100 km (62 mi) into Saturn's atmosphere.

But what about that weird hexagon? Surely that can't be natural. In fact, it is. The most likely explanation is that there's a strong jet stream around the pole that creates a strong gradient in the wind speed. Combine that with rotational forces due to Saturn's short, 10.5-hour day, and you get atmospheric waves that meander up and down in latitude. The same happens to Earth's jet streams, but on Saturn, the waves settle into a stable, six-sided polygon around the pole, as you'll see. Try and time your visit during the northern summer solstice, which happens every 29 years, to see the hexagon change in color from blue to golden as seasonal hazes build up over it.

Address The North Pole of Saturn, Sixth Planet from the Sun, The Solar System | **Getting there** Flybys using Venus, Earth, and Jupiter can provide an efficient route to reach Saturn. | **Tip** If you're lucky, you'll see one of Saturn's megastorms. These erupt at lower latitudes every 20–30 years and are visible for up to six months, leaving side effects in the atmosphere that can last a century.

SOLAR SYSTEM

27 — Olympus Mons
Atop the tallest volcano in the Solar System

It's sunrise and there's a light frost of water ice on the ground. You're standing in the middle of what appears to be a large crater with high cliff walls that rise to 3 km (2 mi) at a distance of 30–40 km (19–25 mi) around you in all directions. But looks can be deceiving, as you're actually in the caldera at the top of Olympus Mons, the largest volcano in the Solar System.

Olympus Mons is located just off the western edge of the Tharsis Plateau on Mars, home to several other enormous shield volcanoes, and everything about it is gargantuan, including the caldera at its summit. Spanning over 600 km (370 mi) in diameter and covering a region roughly the size of Poland, the volcano gradually but inexorably rises to 21 km (13 mi) above the plains around it. That's more than twice the height of Maunakea, Earth's tallest volcano when measured from its ocean floor base. Even the escarpment at the outer edge of Olympus Mons is almost as tall as Mount Everest in places.

To the north and west, the terrain is chaotic, thanks to debris from huge lava-fueled landslides off the volcano, extending as far away as 1,000 km (620 mi). Scientists believe that Olympus Mons was surrounded by an ocean when the first landslides occurred billions of years ago, and others have happened more recently.

One reason Olympus Mons is so oversized is that it has been built progressively over eons. Mars lacks Earth's plate tectonics, so the mountain has stayed fixed over the same hotspot under the martian crust, growing ever larger with every eruption. The lower martian gravity has helped too. The volcano might still be active but dormant these days. Analysis of lava flows on its flanks suggest that they emerged between 115 million and just 2 million years ago, a mere blink of the eye in geological terms. So watch for new activity as you plan your hikes on the giant volcano.

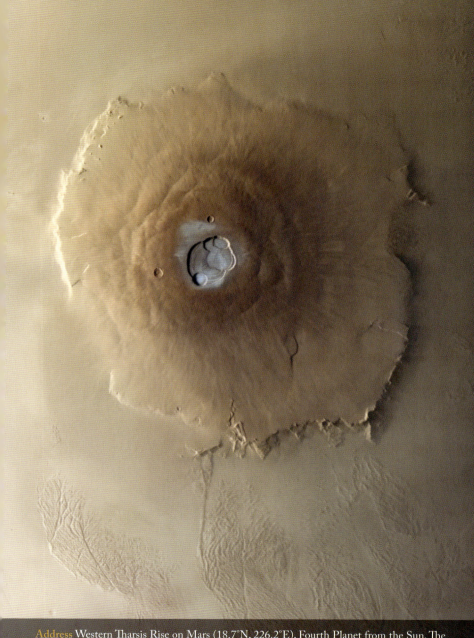

Address Western Tharsis Rise on Mars (18.7°N, 226.2°E), Fourth Planet from the Sun, The Solar System | **Getting there** Mars makes its closest approach to Earth every 26 months at a distance of around 60 million km (38 million mi), so plan your trip accordingly. | **Tip** Arsia Mons is one of the other large shield volcanoes on the Tharsis plateau. In fall, a cloud 1,000 km (621 mi) long trails westwards from its summit.

28 — 'Oumuamua
Our first known interstellar visitor

It'll be a long journey to 'Oumuamua and even longer on the way back because it's moving rapidly away from Earth. But do visit, because this strange object is the first recorded emissary from beyond our Solar System.

Named for the Hawaiian word for "scout" or "first distant messenger," 'Oumuamua was discovered by Robert Weryk in 2017 from Haleakalā Observatory a few weeks after it made its closest approach to the Sun. Astronomers quickly worked out that it was moving so fast that it must have fallen in from interstellar space and will leave the Solar System again soon.

If you do manage to catch up, you'll find that it has some unusual properties. It's reddish in color and changes brightness periodically. That means it must be either elongated and cigar-like in shape or flattened and disk-like, and tumbling. Its size is uncertain though, anywhere from 50 to 1,000 meters (164–3,280 feet) on its largest axis.

Strangely, 'Oumuamua was seen to be accelerating slightly as it headed away from the Sun – objects crawling out of the Sun's gravity well should slow down. Comets expel jets of gas that could cause this, but no tail was seen from 'Oumuamua. As a result, some people suggested that it's an alien spaceship. But there's no real evidence, and almost all astronomers believe 'Oumuamua is a natural object. The best hypothesis is that it's a huge chunk of nitrogen ice like that found on Pluto, thrown into the darkness by a giant collision in a distant planetary system.

We may never know where 'Oumuamua came from, but we do know that it's not alone. There are likely vast numbers of similar objects, ejected violently from orbit around one star, perhaps occasionally briefly encountering others. Another interstellar visitor has been found since, and new space missions aim to intercept one to learn more about these mysterious, void-crossing travelers.

Address The Kuiper Belt, The Outer Solar System (for the time being) | **Getting there** Head towards the constellation of Pegasus, the winged horse, and the sooner you leave, the better. By 2025, 'Oumuamua was well beyond the orbit of Pluto and moving outwards at almost 100,000 km/h (62,000 mph). | **Tip** The brightest star in the general direction of 'Oumuamua's travel is HD223531. Stick with 'Oumuamua, and after some 4.4 million years, the star will glide past two light years to starboard.

SOLAR SYSTEM

29 — The Pale Blue Dot
If you lived here, you'd be home now

Every journey has a beginning, and each one you will take starts on Earth. The same is true of humankind and every living thing we currently know of in the Universe. We all began on this small, rocky planet, orbiting an unremarkable star halfway to the edge of an ordinary galaxy. That's not to say that we are necessarily alone. The vastness of space and its uncountable, star-strewn islands and archipelagos speak against it, especially since we now know that most stars also host planets. It seems likely that we will discover places where life may have arisen in the past or may yet in the future.

But we humans evolved on this planet, intimately connected to its atmosphere and gravity, its ranges in temperature and pressure, and its rivers, oceans, deserts, and mountains. And we are utterly dependent on the vast abundance and variety of life around us, spanning all domains, even if we often fail to return the favor. Of course, we are creative and adaptable – we can, we have, and we will visit other worlds in our Solar System. They promise great adventure and discoveries for a few people, but none of them will prove easy or even possible to settle. The great distances between the stars mean that, in reality, we humans are very likely to live out our future on this planet or close by.

It is a profoundly beautiful planet, a sustaining oasis in the hostile darkness that surrounds us. Certainly, it is groaning under the immense weight of the burdens we have carelessly, greedily, foolishly placed upon it. While we know how to relieve many of those burdens, it remains unclear whether we will act quickly and selflessly enough to do so. It is sobering to think, though, that no matter how badly we treat this planet, there is nowhere else where it will be easier to live.

As Carl Sagan wrote, we must "cherish the pale blue dot, the only home we've ever known."

Address Third Planet from the Sun, The Solar System | **Getting there** Unless you're an astronaut, it's very likely that you're already here. There are so many places to go and things to see, close to home or far away, but there are other guidebooks in this series for that. | **Tip** According to Ancient Greek legends, the center of the world is the Omphalos at Delphi in Greece, the navel of Earth. It's certainly worth a visit to spend some time contemplating our place in the wider Universe.

SOLAR SYSTEM

30 Phobos
A natural space station orbiting the Red Planet

Mars' largest moon Phobos is very different from the moon that orbits Earth. Dark and irregular, Phobos is tiny, only 22 km (14 mi) in diameter. Skimming 6,000 km (3,750 mi) above the planet's surface, it circles Mars three times a day. Once you've touched down on this moon, you'll find spectacular views of Mars looming large overhead, spanning more than a third of the sky, and you'll enjoy many famous martian sights as they roll past.

Asaph Hall III (1829–1907) discovered Phobos in the summer of 1877 from the US Naval Observatory (USNO) in Washington, DC. His wife Angeline Stickney (1830–1892), for whom the largest crater on Phobos is now named, had persuaded him to continue his hitherto fruitless search. Stickney was a mathematician who'd been Hall's geometry and German instructor at college. She'd helped him get his job at the USNO and assisted in his later calculations. She stopped, however, when Hall refused to pay her a man's wage for doing so.

Exploring Phobos, you'll find that it's not a solid body, but rather a porous rubble pile held loosely together by its weak gravity. Many of the small craters on the moon were caused by material that fell back after the big impact that created Stickney. Conversely, the side of Phobos that permanently faces Mars is covered in streaks and striations, which may be evidence of impacts on the planet that ended up sandblasting the moon too. Its surface may also have captured atoms and molecules that escaped the martian atmosphere, preserving a record of the planet's history.

Phobos' origin, however, remains a mystery. Is it a main belt asteroid that was captured by Mars? Or perhaps the aftermath of a giant impact on Mars that threw material into orbit, later coalescing to form the moon? To help answer these questions, scientists are keen to get samples from Phobos, the main aim of the Japanese-led MMX mission, due for launch soon.

Address Orbiting Mars, Fourth Planet from the Sun, The Solar System | **Getting there** The best opportunities for reaching Phobos come every 26 months. Given the moon's low gravity, it's easier to get here than the martian surface, and some propose using Phobos as a staging post for Mars exploration. | **Tip** Deimos is the other moon of Mars and is even smaller, much less cratered, and further out, orbiting every 30.3 hours. Phobos and Deimos are the twin Greek gods of fear and panic.

31 Plumes of Enceladus
Water, water, everywhere, nor any drop to drink

Enceladus isn't just a pretty face – it has hidden depths. With a diameter of 500 km (311 mi), this moon of Saturn was discovered by William Herschel (1738–1822) in 1789 and is the setting for one of the most remarkable detective stories in the history of Solar System exploration.

Early spacecraft visits to Saturn showed that Enceladus was embedded in the planet's puffy E-ring. Scientists immediately wondered whether the moon was making the ring somehow, but the answer had to wait more than two decades until the Cassini mission arrived in 2004 and discovered giant plumes of water vapor gushing from its surface.

Enceladus has a rocky core and an icy crust 30–40 km (19–25 mi) thick. Trapped in between is a global ocean some 26–31 km (16–19 mi) deep. Near the south pole, the crust thins considerably, and water can make its way through deep fissures, known as "tiger stripes," to the surface, where it erupts into space through fumaroles. The water flash-freezes into ice. Some falls back onto the surface of Enceladus, keeping it fresh and making it the shiniest body in the Solar System. The rest of the water trails from the moon and supplies the planet's E-ring.

Every second, 250 kg (550 lbs) of water escapes from Enceladus, and measurements show that it's briny and undrinkable, like the *Ancient Mariner's* ocean. But it also contains a mélange of interesting chemicals, including ammonia, phosphorus, silicates, and simple organic molecules. Those materials might originate from hydrothermal vents on the ocean floor, similar to those found at mid-ocean ridges on Earth, where many scientists believe life may have originated.

Have similar conditions on Enceladus led to the development of life in its ocean? The best way to conduct your investigation is to take a submarine through one of the tiger stripes. Make sure it's armored and has powerful lighting – you never know…

Address Orbiting Saturn, Sixth Planet from the Sun, The Solar System | Getting there Flybys using Venus, Earth, and Jupiter can provide an efficient route to reach Saturn and Enceladus. | Tip Enceladus is in an orbital resonance with the larger moon Dione, providing the heat that drives the geological activity on Enceladus. Dione has long skeins of ice cliffs hundreds of meters high.

SOLAR SYSTEM

32 Pluto
The god of the dark Underworld

After the discovery of Neptune in 1846, astronomers wondered if there was yet another planet beyond it that might also be perturbing the orbit of Uranus. So businessman and astronomer Percival Lowell (1855–1916) built an observatory in Arizona to search for "Planet X," but it wasn't until 1930 that Clyde Tombaugh (1906–1997) found a slowly moving object that seemed to fit the bill. Following a suggestion by 11-year old Venetia Burney (1918–2009), it was named Pluto.

The gateway to the outer Solar System, Pluto has a storied history in the annals of space exploration, culminating in the 2015 flyby of the New Horizons probe. Over time however, it became clear that Pluto wasn't large enough to influence the orbit of Uranus, and refined calculations proved that there was, in fact, no need for a Planet X as envisaged by Lowell. In 2006, the International Astronomical Union reclassified Pluto as a "dwarf planet," part of a family that includes Ceres in the asteroid belt, and also Eris, Haumea, Makemake, and others in the Kuiper Belt beyond Neptune.

Its changed status notwithstanding, Pluto remains a fascinating body. As you approach, a cream-colored world will loom out from the stygian depths, smaller than our Moon. You'll see a frozen surface made of solid nitrogen, with traces of methane and carbon monoxide, and deposits of dark organic molecules, all under a very thin, hazy blue atmosphere. There are windblown dunes made of methane ice "sand," glaciers of nitrogen ice, and tall mountains of water ice, their peaks sprinkled with methane snow. Look for Sputnik Planitia, a bright, smooth plain, perhaps less than 200,000 years old, its surface slowly but constantly overturned by convection. Beneath the surface, it's thought there's a thick water ice crust over a liquid water ocean surrounding a rocky core.

Although our Solar System may only have eight planets again, Pluto is one of many exciting destinations to visit in the Kuiper Belt.

Address The Kuiper Belt, The Solar System | **Getting there** Pluto's orbit is eccentric – it can be as close as 4.4 billion km (2.75 billion mi) to the Sun or as far away as 7.4 billion km (4.6 billion mi). A gravitational slingshot at Jupiter can help you get there quicker if you launch when the planets are optimally aligned. | **Tip** It's worth visiting Hydra, Nyx, Kerberos, and Styx, small irregular moons that might have formed out of the debris made when Pluto and its largest moon Charon collided.

SOLAR SYSTEM

33 Rings of Saturn
Not all those who wander are lost

At first sight, they don't seem real, and the longer you stare at the rings of Saturn, the more outrageously implausible they become. How can they just hang there in space, seemingly pristine and precise, so close to the gas giant planet? Get closer, though, and some of the celestial clockwork that underlies them is revealed.

The rings were discovered in 1609 by Galileo Galilei when he began his telescopic exploration of the Universe. He described them as "ears" on either side of the planet. Decades later, Christiaan Huygens (1629–1695) and Giovanni Domenico Cassini (1625–1712) made improved observations that revealed their true form.

Travel there today, and you'll see that Saturn is orbited by many, many rings and partial arcs of varying brightness, extending from just 7,000 km (4,300 mi) above the planet's atmosphere to 13 million km (8 million mi) away. There are gaps of different sizes between them, often created by the small moons that shepherd them, also creating ripples, wakes, and waves in the rings. Despite their immense width though, the inner rings are generally only a few tens of meters thick, and it has long been known that they are not solid. They're made of particles of almost pure water ice ranging from meters to micrometers in size, continually coming together in clumps and breaking apart again. Some of the fainter, fuzzier, and fatter outer rings are made of material coming from moons themselves, including water ice plumes emerging from Enceladus and dust from others.

Exactly how and when the main rings formed remains unknown. Perhaps they came from a disrupted, icy moon billions of years ago, but some evidence favors a much more recent origin, only a few hundred million years ago. What is certain is that they are slowly disintegrating, raining down on Saturn, and will be gone within a few hundred million more.

Address Orbiting Saturn, Sixth Planet from the Sun, The Solar System | **Getting there** Arriving in orbit around Saturn may involve a circuitous path. The Cassini probe used flybys at Venus, Earth, and Jupiter during its almost seven-year journey to the ringed planet. | **Tip** The small moon Atlas is a must-see. It orbits close to the sharp, outer edge of Saturn's A ring and has swept up material into a huge ridge around its equator, giving it the appearance of a bizarre flying saucer.

SOLAR SYSTEM

34_ The Sun
The star in our back garden

Earth was born along with the Sun 4.6 billion years ago and will die with it when our star expands into a red giant some 5 billion years in the future. For much of human history, this monster in the sky was revered and feared as a god. Today, we know that it is just one star out of hundreds of billions in the Milky Way, albeit with an utterly central role in almost all life on our planet.

With a diameter of 1.4 million km (870,000 mi), the Sun accounts for 99.9 percent of the mass of the Solar System: 950,000 Earths would fit inside if packed as spheres, or 1.3 million if you blended them first. The innermost one percent of the Sun's volume is a nuclear fusion reactor, where the immense density, pressure, and temperature turns 600 million tonnes of hydrogen atoms into helium every second. The liberated energy takes around 100,000 years to percolate up to the 5,500°C (9,900°F) surface of the Sun before crossing space in mere minutes to light, warm, feed, power, and inspire us.

The constancy of this process is reflected in the Sun's impassive face, mostly blank at visible wavelengths, apart from granules and the occasional sunspot. Tune your goggles to x-rays and the ultraviolet to see a roiling surface driving off a constant wind of high-energy plasma. Giant loops and arches betray the presence of an intense magnetic field that can create huge eruptions of material. These are more frequent when the Sun is near the peak of its 11-year cycle. When directed at Earth, the ejected material can cause "space weather" and lead to major challenges for our technological civilization, including damage to satellites and dangerous induced currents in power systems and pipelines.

The desire to get closer to the Sun and learn more about its secrets is something the legends of Icarus, Phaeton, Sampati, and Kua Fu warn us against. Perhaps it's best left to our robot spacecraft, better equipped to handle the Sun's hot and dangerous embrace.

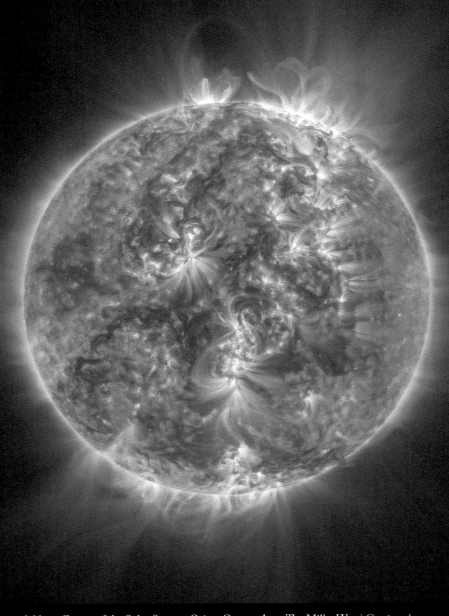

Address Center of the Solar System, Orion-Cygnus Arm, The Milky Way | **Getting there** The Sun is 150 million km (93 million mi) from Earth, 8.3 minutes at the speed of light, although you'll probably want to stop short of that. | **Tip** The heliopause is where the solar wind flowing outwards is balanced by the combined winds of our stellar neighbors. At its closest, it's 18 billion km (11 billion mi) from the Sun. Voyager 1 was the first spacecraft to

SOLAR SYSTEM

35 Sunspots
Tracers of the Sun's internal cycle

As you approach the Sun, the apparently pristine, smooth surface begins to take on a mottled appearance. You'll see cells of hot gas, or "granules," around 1,500 km (940 mi) in diameter, rising and falling every few minutes. They transport energy by convection from below and make the Sun's photosphere glow at about 5,500°C (9,950°F). Then, as the Sun slowly rotates beneath you, a much bigger feature as large as a planet comes into view. This is a sunspot. It has a dark center surrounded by petal-like tendrils, much like a sunflower.

Sunspots are the consequence of strong magnetic fields erupting almost vertically from the surface of the Sun. The central region is known as the "umbra," with temperatures of "only" 2,750–4,250°C (5,000–7,700°F). Although cooler and darker than the surrounding surface, it's all relative. An average-sized sunspot umbra cut out of the Sun and placed in the night sky would glow crimson-orange more brightly than the full Moon. The surrounding "penumbra" is threaded with more horizontal magnetic field lines, and material flows both inwards and outwards along them.

Large sunspots can be seen from Earth with the naked eye when conditions, such as a foggy sky, make it safe to look at the Sun, and astronomers have been cataloging them for thousands of years. Their numbers wax and wane over an 11-year "solar cycle," as the Sun's magnetic poles flip from north to south and vice versa. During some periods in history, such as the 1645–1715 Maunder Minimum, the sunspots almost completely disappeared, reflecting the ever-changing nature of our star.

It's very interesting to watch sunspots grow, merge, and fade away again over days, weeks, and months. Don't get too close though – not only are they extremely hot, but sunspot groups are often associated with other kinds of violent activity, such as dangerous solar flares and coronal mass ejections.

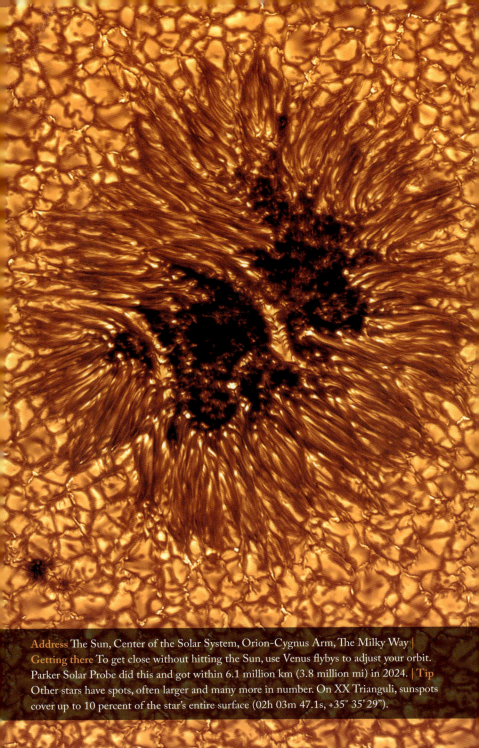

Address The Sun, Center of the Solar System, Orion-Cygnus Arm, The Milky Way | **Getting there** To get close without hitting the Sun, use Venus flybys to adjust your orbit. Parker Solar Probe did this and got within 6.1 million km (3.8 million mi) in 2024. | **Tip** Other stars have spots, often larger and many more in number. On XX Trianguli, sunspots cover up to 10 percent of the star's entire surface (02h 03m 47.1s, +35° 35' 29").

36 — Titan

A smoggy moon with marvels below

There are many moons to visit in our Solar System, from Earth's solitary Luna to the 95 officially recognized ones around Jupiter. But Saturn's largest moon, Titan, is very special – it's the only one to have a substantial atmosphere.

You'll spot the dense nitrogen and methane atmosphere long before you arrive at the 5,150 km (3,200 mi) diameter moon. Ultraviolet light and high-energy particles from the Sun turn carbon-rich molecules into a moon-wide, organic smog that makes Titan look like an orange fuzzball and hides its surface from human eyes. Tune your googles to the infrared, and some of its secrets are revealed. The equatorial regions are home to huge dunes of dark "sand" made of frozen hydrocarbons at temperatures around -170°C (-274°F). Some other dark patches come and go, as seasonal showers of methane and ethane rain soak the ground. These chilly liquids carve river channels and drain in the northern polar regions into seas some 100–200 meters (330–660 feet) deep and as large as the Great Lakes of North America. Some speculate those shallow seas may be home to life very different to that on Earth. There may also be cryovolcanoes spewing a mixture of water, ammonia, and hydrocarbons.

In 1655, Dutch astronomer Christiaan Huygens discovered Titan, and in 2005, the Cassini spacecraft dropped an ESA probe named for him into its atmosphere. Retracing that descent, you also may end up in an alien dry riverbed with a crunchy-soft, *crème brulée*-like surface strewn with hydrocarbon ice pebbles. At ground level, the atmosphere is 60 percent denser than Earth's, but it's nothing you can breathe. Beneath you, there's likely a shell of water ice, a deep water ocean, another shell of high-pressure ice, and a rocky core. It's no wonder scientists are planning to revisit this fascinating world with the NASA Dragonfly robotic rotorcraft, due to arrive in the mid-2030s.

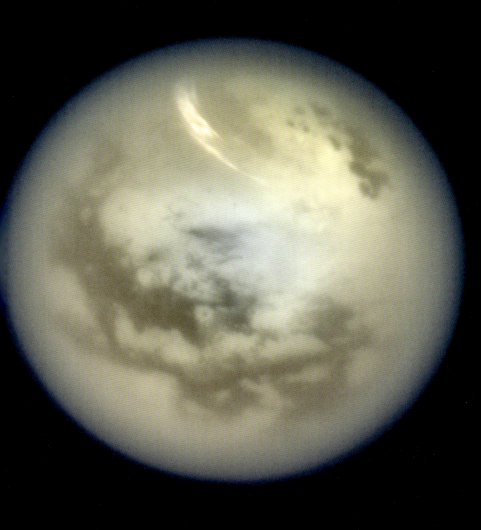

Address Orbiting Saturn, Sixth Planet from the Sun, The Solar System | **Getting there** Flybys using Venus, Earth, and Jupiter can provide an efficient route to reach Saturn. | **Hours** Titan's seasons last about 7 years as Saturn takes 29.5 years to orbit the Sun. If heading for the northern lakes, best visit in winter to avoid summer showers. | **Tip** Iapetus is another of Saturn's moons with one bright, icy hemisphere, the other dark and dusty, and an equatorial ridge 20 km (12.4 mi) high.

SOLAR SYSTEM

37 Triton
A smörgåsbord of icy delights

Orbiting Neptune far from the Sun, Triton's daylight is almost a thousand times fainter than on Earth. However, despite the absolutely frigid temperatures of -235°C (-391°F), Triton is no dead moon – indeed, it is a dynamic world of ice in many forms, and there's much to be discovered in one of the most geologically active locations in the Solar System.

Triton was discovered by William Lassell (1799–1880) in 1846. As you approach, you'll notice that it's very reflective, thanks to an icy crust made of frozen water, nitrogen, methane, and ammonia. That's partly why it's so cold – it doesn't absorb much of the meager sunlight falling on it. But as you get closer, you'll see that the ice comes in many different varieties. The south pole is rugged, while, closer to the equator, there is old and wrinkled "cantaloupe terrain." There are large, smoother plains and frozen lakes with almost no impact craters. These were made as liquid water flowed from cryovolcanoes within the past hundred million years, the blink of an eye in geological terms. Then there are cracks, ridges, and valleys across the moon likely due to a freeze/thaw cycle. You'll see long-lived, geyser-like plumes of gas and dust rising up to 8 km (5 mi) above the surface. Whether the warmth that powers them comes from the absorption of sunlight or residual heat in a subsurface water ocean is something scientists are very keen to discover.

After ascending back through Triton's tenuous nitrogen atmosphere, you'll notice that it's much larger than all of Neptune's other moons. Weirdly, it circles in the opposite direction to the planet's rotation and in an orbit tilted relative to the equator. That points to Triton having once been a separate dwarf planet, larger than Pluto, unluckily captured by Neptune's gravity long ago. Only 40 percent of Triton was photographed during the 1986 Voyager 2 flyby, leaving many reasons for you to visit.

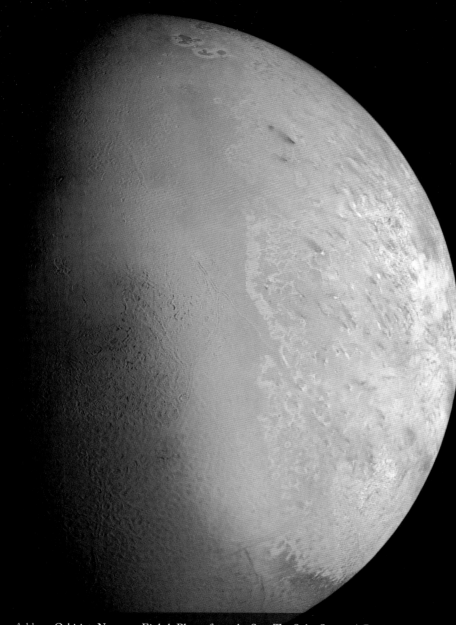

Address Orbiting Neptune, Eighth Planet from the Sun, The Solar System | **Getting there** Gravitational slingshots at Jupiter, Saturn, and Uranus can speed up your journey if you launch when the planets are best aligned. | **Tip** The smaller moon Nereid is also worth visiting, as it may have been knocked into its highly eccentric orbit when Triton was captured. It has never been seen close-up, and even its shape is uncertain.

38 Uranus
An enigmatic, sideways world

Far beyond the orbit of Saturn, you will encounter the first planet discovered since the invention of the telescope. The bland appearance of Uranus belies its hidden depths and storied history.

Barely visible to the naked eye, Uranus had been cataloged as a faint star for almost two millennia before William Herschel looked through his telescope in 1781 and saw that it was extended. He thought it was a comet initially, but once it was confirmed as a planet, he proposed calling it Georgium Sidus after his patron King George III. This was not a universally popular choice, and ultimately the suggestion by Johann Elert Bode (1747–1826) to name it for the Greek god of the sky and grandfather of Zeus won the day. Interestingly, the name of the newly-discovered element uranium was also inspired by Bode's proposition.

The cold atmosphere of Uranus is mostly hydrogen and helium, with traces of methane, giving it its signature pale green color. As you descend, it gets warmer and the pressure rises, the atmosphere slowly changing into a hot dense liquid of methane, water, and ammonia, molecules found as ices when the Solar System was young. Strangely, the planet lies on its side, perhaps due to an ancient collision. This orientation results in bizarre seasons, where the northern and southern hemispheres experience 21 years of constant daylight in summer and 21 years of darkness in winter, interspersed with "normal" springs and falls. It is circled by 13 faint ice rings and 28 moons, some of which orbit in the opposite direction to the planet's rotation, suggesting that they're likely captured asteroids.

Uranus has only been briefly visited by one spacecraft, Voyager 2 in 1986. Since then, astronomers have discovered many planets orbiting other stars that seem to share its "ice giant" properties. So you'll have much to study and learn during your visit.

Address Seventh Planet from the Sun, The Solar System | **Getting there** Gravitational slingshots at Jupiter and Saturn can help you on your trajectory to Uranus if you launch when the planets are well aligned. | **Tip** Stop off at Titania, the largest moon of Uranus. Roughly equal parts rock and ice, its relatively smooth surface is split by giant faults and deep canyons, making for interesting hikes.

SOLAR SYSTEM

39 Venus
Our deadly planetary neighbor

They say that fortune favors the bold, but you'll have to be exceptionally brave to visit the surface of Venus – and remarkably lucky to live to tell the tale. It was long thought that our nearest planetary neighbor was perhaps the best chance we had of finding life elsewhere in our Solar System. It is not.

Venus is a little smaller than Earth, orbits the Sun in 225 days, and lies an average of 45 million km (28 million mi) sunward from us. As you approach, Venus won't give up many of its secrets though. It's shrouded in dense, white clouds from pole to pole, and there is no sign of the surface. But tune your goggles to the ultraviolet, and you'll see clouds whipping around the equator at 300 km/h (185 mph). When you reach those clouds about 74 km (46 mi) above the surface, you'll find that they're actually a fog of thick, corrosive, sulfuric acid. Claims that some form of alien organism may be living in the clouds remain controversial.

Descend another 40 km (25 mi), and the view suddenly clears. The cloud above you blocks much of the light even though you're relatively close to the Sun, but outside pressures and temperatures are almost comfortable here. Below, you'll see a fractured surface with giant shield volcanoes, highlands, and mountains. There are no seas or oceans, as Venus lost all of its water billions of years ago. As you continue your descent, conditions become inexorably harsher. The atmosphere gets so thick, you won't even need a parachute to slow you down. By the time you touch down, the temperature will have reached 460°C (860°F), and the pressure is the same as being almost a kilometer (0.6 mi) deep in Earth's oceans. The planet's carbon dioxide atmosphere and the greenhouse effect have turned it into a deadly oven.

Leave any actual exploration to your heavily armored robots, though even the sturdiest of them have lasted only a few hours on the surface of Venus.

Address Second Planet from the Sun, The Solar System | **Getting there** If you set off when Venus is best aligned with Earth, you can reach it in just a few months. If you've decided that you'd like to orbit the planet rather than just go straight to the surface, it'll take a little longer. | **Tip** Venus doesn't have a regular moon like Earth does. Rather, Zoozve, a small asteroid discovered in 2002, makes huge loops around the planet as they both orbit the Sun. Definitely worth a side trip.

40 AG Carinae
Getting ready for its grand finale

"Live fast, die young." This epithet is not only associated with rock stars, but also with actual massive stars burning through their fuel at prodigious rates. As these stars near the end of their lives, the struggle between gravity pulling inwards and the light pressure pushing outwards yields dramatic and unpredictable changes in size, temperature, and brightness, and they also shed vast amounts of their outer envelope. These are "luminous blue variables" (LBVs) and they're so bright, they can even be seen in nearby galaxies from Earth. But this phase of their life is short, making LBVs rare – fewer than 100 are presently known. When the fuel finally runs out, they collapse dramatically and rebound in a supernova explosion.

AG Carinae is an LBV, and that'll become obvious long before you arrive. It is one of the brightest stars in the Milky Way, and everything about it is impressive. It started life just a few million years ago with around 100 times the mass of our Sun and has since shed up to half of that. You'll see it pulsating dramatically and irregularly on decades-long timescales. Even at its minimum size, it's 50 times larger and 1.5 million times more luminous than the Sun, at a temperature of 20,000°C (36,000°F). As it pulsates, you'll see it swell to 10 times that size, although perhaps counter-intuitively, it gets 30 percent fainter, because it also cools to 8,000°C (14,400°F) as it expands.

Some 10,000 years ago, AG Carinae ejected 20 percent of its mass into a large envelope now lit by the star. The star's million km/h (620,000 mph) wind has cleared a cavity spanning almost 30 light years and continues to sculpt intricate, dusty clumps and streamers around its edge. Exactly when AG Carinae will explode in the next few million years is unknown. But you should consult your insurance company before traveling there.

Address Fourth Galactic Quadrant, The Milky Way (10h 56m 11.8s, -60° 27' 13") | **Getting there** Head towards the constellation of Carina, the ship's keel, and travel for 17,000 years at the speed of light, although the exact distance remains controversial. | **Tip** En route to AG Carinae, make a stop at NGC3496, a cluster with one group of stars about 400 million years old and another at 600–900 million years (10h 59m 30.5s, -60° 20' 06').

41 Alpha Centauri System
Our nearest stellar neighbor(s)

After exploring the many exciting destinations in our Solar System, you may be ready to venture into the Milky Way. The distances get much larger and the journeys longer, so start off gradually with a visit to Alpha Centauri, the nearest star to the Sun. Well, sort of.

Alpha Centauri is a triple star system. Both A and B components, known as Rigel Kentaurus and Toliman, respectively, are similar in mass and luminosity to our Sun. They swing around each other every 80 years, their mutual orbit covering roughly the size of our Solar System. Combined, they make the third brightest star in our night sky, but their sibling, component C, is much fainter. It's a tiny, red dwarf star, and even in photographs, it's hard to see. It lies more than 0.2 light years from A and B and orbits them every 550,000 years. It is currently the closest star of the three to Earth, hence its name Proxima Centauri.

Therefore, it's best to see Proxima Centauri first, and it's the most interesting as well. You'll find a dim star orbited by at least one planet and perhaps three. The confirmed exoplanet, Proxima b, has roughly the same mass as Earth, but it's not yet known whether it's dense and rocky like Mercury or much fluffier. Its distance from Proxima Centauri leads to temperatures suitable for liquid water, but don't make plans to camp here. The star unleashes huge flares of high-energy particles that would strip away any atmosphere the planet had and sterilize its surface. The star is parsimonious with its nuclear fuel, though, and will live for another four trillion years.

Head to Rigel Kentaurus next, and you may find a planet in its habitable zone. But if so, it's Neptune-like and thus not very camping-friendly either. No planets have been discovered orbiting Toliman to date, but that may have changed by the time you arrive, as many other adventurers are likely to explore the Alpha Centauri system.

Address Fourth Galactic Quadrant, The Milky Way (14h 39m 36.5s, -60° 50' 02") | **Getting there** Head towards the constellation of Centaurus, the centaur, and travel for 4.25 years at the speed of light. Check the exact coordinates beforehand, as Alpha Centauri moves quickly across the sky. | **Tip** Beta Centauri, or Hadar, is a triple system. It's much further away than Alpha Centauri but very bright. The |Xam Bushmen of South Africa see them together as two men turned into lions (14h 03m 49.4s, -60° 22' 23").

42 Betelgeuse
All the colors of the stars

Expect the unexpected when you travel to Betelgeuse, although, contrary to common media hype, this star most likely will not explode during your trip – that's not expected for at least 100,000 years.

One of the brightest stars in the constellation of Orion, Betelgeuse, or Alpha Orionis, is much cooler than the other prominent members at about 3,400°C (6,200°F). You can see its ruddy orange color with the naked eye, while the other bright stars are blue-white. Betelgeuse started its life as a member of the Orion star-forming complex some 10–12 million years ago and likely had a close companion. When that companion exploded as a supernova, Betelgeuse was released from their mutual orbit, and it's now considerably closer to Earth. Betelgeuse too is also nearing the end of its life, having turned from a hot OB star into a red supergiant, and when you get there, you'll be astounded. It's hugely bright and bloated, around 75,000 times the luminosity and 700 times the diameter of our Sun. If it were in our Solar System, it'd extend beyond the orbit of Mars. There's no well-defined outer edge though – it's constantly roiling with massive sunspots, convection cells, and giant plumes. Its cool atmosphere ejects large clouds of dust, and occasionally they cross in front of Betelgeuse, significantly dimming it for years at a time.

The Arabic name means "the hand of the giant," i.e. Orion, and the brightness, variability, and deep orange color of Betelgeuse have given it a prominent place in many other myths around the world. In modern culture, it inspired the name of Tim Burton's 1988 film *Beetlejuice*. In *The Hitchhiker's Guide to the Galaxy* by Douglas Adams (1952–2001), the character Ford Prefect is said to come from its vicinity. His home planet likely won't survive when Betelgeuse does explode as a supernova, but again counter to common tropes, it probably won't endanger Earth.

Address Third Galactic Quadrant, The Milky Way (05h 55m 10.3s, +07° 24' 25") | **Getting there** Head towards the constellation of Orion, the hunter, and travel for 400–550 years at the speed of light – the distance to Betelgeuse is surprisingly uncertain. | **Tip** Visit Meissa, or Lambda Orionis, for contrast. One of the hottest stars in Orion at 36,000°C (65,000°F), it's relatively faint at visible wavelengths as most of its light is emitted in the ultraviolet. It's also a multiple system (05h 35m 08.3s, +09° 56' 03").

43 _ Boomerang Nebula
Baby, it's cold outside

From your extensive travels through space, you'll have realized just how cold it can get out there. Far from any source of heat, like a star or a planet, the temperature bottoms out at a very chilly -270.4°C (-454.8°F), just 2.7°C (4.9°F) above absolute zero. The origin of that paltry warmth is the Cosmic Microwave Background (CMB), the left-over echo of the Big Bang, and it pervades the whole of space. Nothing can be colder.

Or can it? If you want to experience the very lowest temperatures available outside of a lab, then you should head towards the Boomerang Nebula, named by astronomers who discovered it from Australia in the 1970s. As the nebula comes into view, you'll see a faint, hourglass structure pinched to a narrow waist around a star. You might think that this is a young star surrounded by a circumstellar disk, with light escaping from the poles and illuminating its dusty birth cloud. In fact, it's a dying red giant in the process of expelling its outer envelope to form a pre-planetary nebula. A stellar wind pushes out material in a bipolar shape that is then lit up by the shrinking star. Later, once the star has turned into a white dwarf, it will be hot enough to ionize the surrounding shell of gas and form a planetary nebula. But for now, that rapidly growing shell is the key to a very interesting physics trick.

As the gas expands outwards at about 600,000 km/h (370,000 mph), it loses heat, getting progressively colder and colder. This is the same way a refrigerator works, by compressing and expanding a gas, but on an altogether vaster scale. And unlike your domestic fridge, observations at millimeter wavelengths show that temperatures here reach a spectacular low of just 1°C (1.8°F) above absolute zero, well below the CMB. Once you return home from the Boomerang Nebula, you'll truly be able to say you've been to one of the coolest places in the Universe.

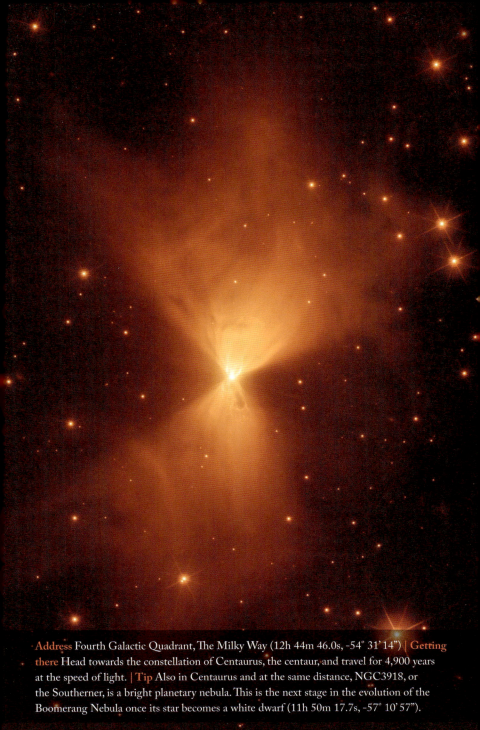

Address Fourth Galactic Quadrant, The Milky Way (12h 44m 46.0s, -54° 31' 14") | **Getting there** Head towards the constellation of Centaurus, the centaur, and travel for 4,900 years at the speed of light. | **Tip** Also in Centaurus and at the same distance, NGC3918, or the Southerner, is a bright planetary nebula. This is the next stage in the evolution of the Boomerang Nebula once its star becomes a white dwarf (11h 50m 17.7s, -57° 10' 57").

MILKY WAY

44 Bubble Nebula
Grace under pressure

Some places in space seem beautifully serene – at first. But take a closer look, and the struggle among cosmic forces on a vast scale becomes apparent. Take, for example, the Bubble Nebula, also known as NGC7635, first discovered in 1787 by William Herschel.

From a distance, you'll see an almost-circular structure about 7.5 light years wide and outlined in a tracery of blues and greens, signifying the presence of ionized oxygen and hydrogen in this instance. This structure is being blown by a wind moving at around 7.6 million km/h (4.7 million mph) away from a very hot, young star about 44 times as massive as our Sun. But as you continue to get closer, you'll see that the star in question, BD+60°2522, lies far from the center of the bubble. Is the star moving? Or is its wind stronger on one side than the other?

The answer lies in the wider environment. As you look around the Bubble Nebula, you'll see that it's accompanied by a giant, molecular cloud of gas and dust, material left over from the formation of BD+60°2522 and other young stars. As the bubble expands around the star, it meets the surrounding cloud, and they collide supersonically. Because the cloud is denser to one side of the massive star, the bubble is more constrained on that side and freer to expand on the other, leading to the asymmetry. However, the intense, ultraviolet light from the star is able to cross the pressure balance boundary and heats up the surrounding cloud, causing it to glow in reds and yellows. You can see a column of this material apparently close to the star, where new stars may be forming inside.

This lovely scene is only temporary, so you should visit soon. The bubble may be only 40,000 years old and could evolve rapidly, while BD+60°2522 is burning through its nuclear fuel furiously and will likely collapse then explode as a supernova within the next 10 million years.

Address Second Galactic Quadrant, Perseus Arm, The Milky Way (23h 20m 40.1s, +61° 10' 52") | **Getting there** Head towards the constellation of Cassiopeia, the mythical queen of Aethiopia, and travel for 8,200 years at the speed of light. | **Tip** About halfway to the Bubble Nebula, you can visit the Scorpion Cluster, also known as M52 or NGC7654. It's about 120 million years old and has thousands of members. Take a good look as you pass by (23h 24m 45.4s, +61° 35' 22").

45 Carina's Bok Globules
Islands in the storm

When you visit a region where young stars are being born, your eye will likely be drawn to the new stars themselves, their protoplanetary disks, and their jets and outflows. It's often a lot harder to see the clouds of material from which the stars are made, at least at visible wavelengths. Cold and replete with dust, these clouds are usually dark compared to their surroundings. But in some places, the lighting conditions can conspire to enhance their visibility. The dramatic maelstrom of the Carina Nebula is one such location.

As you arrive at the region, the bright, ionized nebulosity and dense stellar cluster nestled in it will surely grab your attention initially. But look between the stars. You'll see many small, ragged blobs silhouetted against the nebula. These are known as Bok globules after astronomer Bart Bok (1906–1983), who, working with Edith Reilly (1917–1988), discovered many of them in the 1940s while studying photographs of star-forming regions. They're full of cold, dense gas and dust, the latter blocking the background light. Bok and Reilly suggested that they might be a kind of cocoon, a place where new stars will condense under gravity's pull. If you switch to infrared vision, the dust becomes partly transparent, and you can see through the globules. Some will have young stars inside, often in multiples.

The globules in the Carina Nebula are particularly interesting because they're in a race against time. Remnants of the original cloud from which the region formed, a few of the smaller globules are dark, but many, including the largest, nicknamed the "Caterpillar," are glowing around the edges. This glow is a signpost of the impact of the intense, ultraviolet light from nearby massive stars heating and eroding the globules. The larger ones may survive this onslaught for long enough to make stars, but many of the rest will simply shrivel and disappear.

Address Carina Nebula, Fourth Galactic Quadrant, The Milky Way (10h 44m 32.3s, -59° 34' 38") | **Getting there** Head towards the constellation of Carina, the ship's keel, and travel for 7,500 years at the speed of light. | **Tip** Very close by is Eta Carinae, a remarkable object. A pair of massive stars in a compact egg timer-shaped nebula, its brightness has varied dramatically over the past two centuries. The larger star could explode as a supernova any day, so plan to observe from a safe distance (10h 45m 03.6s, -59° 41' 04").

46 Cat's Eye Nebula
The beginning of the end for a star

Stars spend most of their lives fusing hydrogen into helium in their cores. The energy produced supports them against the inward tug of gravity. Stars in this steady-state phase are said to be "on the main sequence." But all good things come to an end, and when the supply of hydrogen in the core starts to run out, the star evolves as the balance of energy, gravity, and fusion products changes. How long the main sequence phase lasts and what happens after depends on how massive the star is. For stars with more than eight times the mass of the Sun, that evolution ends in a supernova. A visit to the Cat's Eye Nebula will illustrate the fate awaiting lower-mass stars.

You'll see a hot, white star surrounded by shells and bubbles in pink, blue, and orange. As the star left the main sequence, its core contracted, while the outer parts expanded, blowing away up to 50 percent of the star's mass in its wind. The star continued to shrink and its temperature continued to rise, until it was hot enough to ionize the ejected gas and turn it into a planetary nebula.

William Herschel discovered the Cat's Eye Nebula in 1786, and William Huggins (1824–1910) later showed it to be mostly ionized hydrogen and helium. Within its broadly spherical shape is a complex structure of expanding shells, knots, arcs, and jets that reflect the presence of fast winds, an accretion disk, and perhaps a binary companion to the central star. That star is now at 100,000°C (180,000°F) and is surrounded by even hotter gas emitting x-rays.

But you'll need to visit soon, as this beautiful phase in the dying star's life is ephemeral, lasting as little as 10,000 years. The star will cool and no longer be able to ionize the nebula, so it will go dark. But the material, enriched with the heavier element ashes of nuclear fusion, will spread back into the galaxy and become part of the raw material for making new stars, a form of cosmic recycling.

Address Second Galactic Quadrant, The Milky Way (17h 58m 33.4s, +66° 38′ 00″) | **Getting there** Head towards the constellation of Draco, the dragon, and travel for 3,300 years at the speed of light. | **Tip** The Cat's Eye Nebula lies very close to the North Ecliptic Pole, the point in the sky perpendicular to the plane of Earth's orbit around the Sun (18h 00m 00.0s, +66° 33′ 39″).

MILKY WAY

47 — Cederblad 110
Expect a frosty reception

Visiting vast, crowded clusters of stars as they're being born from giant turbulent clouds of gas and dust can get exhausting. So a trip somewhere quieter, where things proceed more calmly, might offer you some respite. The Chamaeleon dark clouds in the far southern sky would be perfect. Young stars here are found in small, low-density clusters without noisy, massive stars and their strong winds and ionizing photons.

As you approach the Chamaeleon I cloud, you'll see three patches of blue nebulosity shining in the dark dust, each lit up by young stars. One of these nebulae is called Cederblad 110. If you switch your googles to infrared as you get closer, you'll see a small group of stars embedded in the dusty cloud, surrounded by wisps and whorls in blue and orange. One interesting object here is IRS4, thought to be a pair of very young protostars. Less than a million years old, IRS4 is circled by a disk of gas and dust, perhaps forming new planets. Light escaping above and below the disk illuminates fans of blue-white nebulosity.

Dust particles reflecting the light of IRS4 and another nearby star give Cederblad 110 its diffuse, blue glow. Winds and maybe magnetic fields shape its delicate striations and ripples. You'll also see many faint, orange sources in the region, but most are background stars and distant galaxies seen through the thin dust cloud. If you analyze the light of one of those background stars, you'll find dips imprinted on its spectrum at certain wavelengths thanks to absorption by ices frozen on dust grains in the cloud. These ices include regular water, but also ammonia, methane, carbon monoxide, carbon dioxide, and organic compounds, such as methanol and perhaps ethanol and acetone too. As these icy grains may one day grow into pebbles and then planets, it's fascinating to find some of the basic ingredients for making life already in place.

Address Fourth Galactic Quadrant, The Milky Way (11h 06m 44.8s, -77° 23' 03") | **Getting there** Head towards the constellation of Chamaeleon, the color-shifting lizard, and travel for 620 years at the speed of light. | **Tip** The young, bright star HD97048 illuminating the nearby Cederblad 111 nebula is surrounded by a disk of gas and dust with gaps in it, probably carved by young protoplanets (11h 08m 03.4s, -77° 39' 17").

48 Cometary Globule 4
Art imitating life?

In *The Empire Strikes Back* (1980), the fifth episode in the *Star Wars* saga, the Millennium Falcon is almost eaten by a giant space slug living inside an asteroid. But sometimes, real life can be much stranger than fiction, as you'll realize when you arrive at Cometary Globule 4 (CG4). Well, sort of.

CG4 is located in the Gum Nebula, a huge, faint H$_{II}$ region linked to the formation of several clusters of stars around 30 million years ago. As you look around the edges of the glowing nebula, you'll see many dark clouds of gas and dust with faintly luminous tails pointing away from the center. Their appearance gives rise to the name "cometary globule," yet another example of confusing nomenclature in astronomy.

The most striking among them is CG4, colloquially known as "God's Hand," discovered on photographic plates in 1976 by astronomers Tim Hawarden (1943–2009) and Peter Brand. The column of gas and dust is eight light years tall and ends in a gaping maw that seems to be reaching for a silver disk bearing an uncanny resemblance to Han Solo's ship. But that's where the fanciful coincidence ends. You're actually seeing an edge-on spiral galaxy, millions of light years in the background. Also, rather than reaching for the galaxy, CG4 is slowly shrinking back, eroded by the strong winds and intense, ultraviolet light of nearby hot, young stars that also illuminate it.

These stars and associated young clusters are thought to have formed more recently after the first burst of star formation in the region compressed surrounding material. In the same vein, it's thought the cometary globules are remnants of the original molecular cloud shredded by that initial activity. Under the influence of today's massive stars, they too may yet go on to make more, and indeed, some globules in the region have Herbig-Haro jets emerging from them, a tell-tale sign of young stars inside.

Address Gum Nebula, Third Galactic Quadrant, The Milky Way (07h 34m 05.4s, -46° 56' 59") | **Getting there** Head towards the constellation of Puppis, the stern of Argo Navis, for around 1,300 years at the speed of light. | **Tip** Zeta Puppis is one of the stars illuminating CG4. It is a blue supergiant and a fast-moving runaway star, probably ejected from a binary system when its companion exploded as a supernova (08h 03m 35.1s, -40° 00' 12").

49 Cosmic Bat Nebula

Appearances can be deceptive

You might want to draw the curtains over the forward-looking windows as you fly towards this destination. Otherwise, the huge bat flying towards *you* might lead to nightmares.

A dark cloud with the prosaic catalogue number Lynds Dark Nebula 43 (LDN43) gives us one of the finest examples of pareidolia, where humans see familiar patterns where there actually are none. In the case of LDN43, it's easy to see why astronomers have nicknamed it the "Cosmic Bat Nebula." Resembling a brown-furred flying fox with widely-spread wings, long ears, and rather demonic eyes and mouth, it makes a terrifying, vampiric impression.

The bat is just an illusion drawn by the feverish human mind, but the reality of LDN43 is no less interesting, and it is well worth a visit. The cloud is made of dense gas and dust, which screen out background stars and nebulosity, giving the bat its outline. Such clouds are the birthplaces of new, low-mass stars. The brightly-lit region of the bat's mouth and chin betray the presence of just such a young protostar buried in dust, some of its light escaping and illuminating the surroundings. This object is called RNO91 for "red and nebulous" and is less than a million years old. Once you arrive, you'll find it circled by a swirling disk of material that may be forming new planets, while also blowing other material away in an outflow that creates the bat's ears. The bright patch under one of the bat's wings is caused by another star illuminating the dusty surroundings of LDN43, although this one is older, between two and six million years old.

Other instances of pareidolia in astronomy include the Man in the Moon and the Face on Mars. Here on Earth, people see animals and faces in the clouds, as Hamlet and Polonius discuss in Shakespeare's eponymous play. There are many other examples in art, architecture, and music, some accidental, some deliberate.

Address Ophiuchus Dark Clouds, First Galactic Quadrant, The Milky Way (16h 34m 29s, -15° 47' 01") | **Getting there** Head towards the constellation of Ophiuchus, the serpent bearer, and travel for 400 years at the speed of light. | **Tip** About halfway to LDN43, you'll encounter Phi Ophiuchi, a giant star a little cooler than the Sun but over 100 times as luminous. It has run out of hydrogen fuel and is nearing the end of its life, so plan your visit accordingly (16h 31m 08s, -16° 36' 46").

50 Crab Nebula
A millennium-old stellar explosion

As you head towards the Crab Nebula, you'll see a tangled web of glowing gas, dust, and plasma. William Parsons, 3rd Earl of Rosse (1800–1867) saw a resemblance to a *decapod brachyuran* in its filaments, hence the name. But this is no crustacean. It's the remains of a dead star.

Shortly before dawn on a late spring day in 1054, Chinese astronomers noted the appearance of a "guest star" in the eastern sky. The star increased in brightness over a couple of months until it could even be seen during daytime. It remained visible at night for almost two years. Only centuries later was this remarkable phenomenon identified as a supernova. It happened when a star perhaps 10 times the mass of our Sun ran out of fuel. With no energy to support it against gravity, the core collapsed. Moments later, the star exploded.

Visiting today, you'll see material expanding away from the immense explosion at almost 5.5 million km/h (3.4 million mph), roughly 0.5 percent of the speed of light. It should have slowed more by now, but the debris is threaded with a strong magnetic field that is channeling some additional outward energy. The source of that energy is a neutron star, the remaining 20 percent of the original star compressed into a tiny, city-sized ball. This neutron star is also a pulsar, rotating furiously 30 times a second and sweeping a lighthouse-like beam around the sky. As the pulsar slows, the lost energy feeds the nebula. The rotation rate also occasionally "glitches" as the interior rearranges itself in titanic starquakes.

The Crab Nebula and its pulsar glow at all manner of wavelengths. By sweeping your goggles up and down the electromagnetic spectrum, you'll be able to distinguish emission from the various atoms, molecules, and electrons. The core around the pulsar is very dynamic, changing on timescales of days, so no two visitors will see exactly the same nebula.

Address Third Galactic Quadrant, The Perseus Arm, The Milky Way (05h 34m 31.9s, +22° 00' 52") | **Getting there** Head towards the constellation of Taurus, the bull, and travel for 6,200 years at the speed of light. | **Tip** By adding one percent to your journey time to the Crab Nebula, you can detour via RNO54, a weird nebula wrapped around a rare type of erupting young star called an FU Ori (05h 42m 21.2s, +22° 36' 47").

MILKY WAY

51 CW Leonis
Ashes to ashes, dust to dust

After millions of years of stable fusion burning on the main sequence, changes come fast when a star runs out of its nuclear fuel, which can involve stars shedding excess weight en route to reinventing themselves, as you'll see when you reach CW Leonis, or IRC+10216.

Eric Becklin and collaborators discovered CW Leonis in 1969 while making one of the first ever infrared surveys of the sky, using a 1.6 meter (62 inch) diameter telescope made from epoxy resin. Almost all of the visible wavelength light from the star is scattered and absorbed by dust surrounding it. The dust then heats up, and CW Leonis positively blazes in the infrared. The star itself is the source of that dust. After leaving the main sequence, CW Leonis swelled into a red giant star, pulsating and varying in brightness by a factor of six every 21 months. It is making huge amounts of carbon-rich dust in its cool atmosphere, that is then blown gently away by the star's wind. As much as half of the star's original mass has been lost this way over the past 100,000 years or so.

You'll see a series of delicately sculpted shells around the star. The structures likely reflect periodic changes in the rate at which the star is losing dust, linked to a magnetic cycle within. There are hints that CW Leonis may have a close companion that could also be playing a role, so make sure to look out for that.

And be sure to have a look back at the system as you leave. Farther out from the star and the brighter inner dust shells, you'll see a faint, semi-circular arc about 2.7 light years across against the darkness of space. This is being created as CW Leonis ploughs through interstellar space at speeds of around 330,000 km/h (205,000 mph), sweeping hot gas into a bow shock in front of it and then into a turbulent tail behind. Not long from now in cosmic terms, CW Leonis will move to the next stage of its life as its remaining core transforms into a tiny, dense white dwarf.

Address Third Galactic Quadrant, The Milky Way (09h 47m 57.4s, +13° 16' 44") | **Getting there** Head towards the constellation of Leo, the lion, and travel for 400 years at the speed of light. | **Tip** Stop off at Regulus, the brightest star in Leo. It's a multiple system comprising two pairs of stars and perhaps a more distant brown dwarf. The brightest star is rotating so rapidly that it's highly oblate, making it brighter at the poles than at the equator (10h 08m 22.3s, +11° 58' 02").

MILKY WAY

52 Cygnus X
The spot for students of star formation

Cygnus, one of the best-known constellations in the northern summer sky, is home to your next destination. You can see hints of it along the Milky Way in the bright North America Nebula near Deneb, the swan's tail, and other nebulosity surrounding Sadr, the heart. But you'll be heading into the dark between them, a huge dust cloud called the Cygnus Rift. Beyond it lies the largest known star-forming region in our part of the galaxy, Cygnus X. Switch your goggles to infrared to cut through the dust and get a spectacular preview.

The core of Cygnus X is a giant collection of molecular clouds full of cold, dense gas and dust, totaling around three million times the mass of our Sun in a region 650 light years across. Many of them are actively making huge numbers of young stars of all masses in hundreds of individual regions and several large groups called "associations." There are many protostars, too, yet to emerge from their dark cocoons.

In turn, these stars are pushing back on their natal material through ultraviolet light, winds, jets, and outflows, sculpting the clouds around them. There's evidence for supernova explosions ripping them apart in places. You'll have seen the same effects working in other regions, but everything's on an altogether larger scale in Cygnus X. Some individual star-forming regions, such as DR21, are massive in their own right but almost lost in the turbulent fire of the complex.

How many stars do complexes like this produce? How many form in dense clusters around massive ones rather than distributed more widely? And how does their powerful feedback end one generation of star formation by blowing away clouds, while perhaps kickstarting another by compressing them? Because Cygnus X is fairly close by, you can use your findings here to help understand what's going on in more distant regions of the Milky Way and other galaxies beyond.

Address First Galactic Quadrant, Orion-Cygnus Arm, The Milky Way (20h 30m 03.4s, +40° 58' 14") | **Getting there** Head towards the constellation of Cygnus, the swan, and travel for 5,500 years at the speed of light. | **Tip** The similarly-named Cygnus X-1 lies in the outskirts of the complex and is one of the brightest x-ray sources in the sky. It was the first source that astronomers agreed was a black hole. It's a must-visit, especially for fans of Canadian progressive rock band Rush (19h 58m 21.7s, +35° 12' 06").

53 — Elephant's Trunk Nebula
Reaching for the stars

Humans have long since been obsessed with building ever taller structures, from pyramids and ziggurats, cathedrals and mosques, to today's skyscrapers with heights flirting with 1,000 meters (3,280 feet). But even the proposed space elevator reaching 100,000 km (62,000 mi) above Earth's equator would pale in the face of true space "towers" such as the Elephant's Trunk Nebula.

On arrival, you'll see a slightly wonky "spacescaper" made of gas and dust being sculpted by ultraviolet light emitted by a cluster of hot, massive, young stars known as Trumpler 37 in the center of the IC1396 star-forming complex. After those stars were born, they started to clear away the remnants of their natal molecular cloud. But some parts of the cloud were denser than others, and light-driven erosion preferentially removed the less-dense surroundings, leaving a clump standing at the top of a thinner pillar. The same effect is seen on Earth in eroded rock structures, including mesas, buttes, and hoodoos.

There are several such pillars in IC1396, and the Elephant's Trunk is the most prominent, as its head is strongly illuminated by a massive star located nearby. The pressure of that starlight has also compressed the clump, and new stars are being born inside as a result, some of them excavating a small cavity in the head.

The scale of the pinnacle below the head is almost unimaginable – from the top to its base in a rim of dark clouds, it spans around 200 trillion km (124 trillion mi). If you stood at the top and flashed a laser beam, it would take 21 years for someone at the bottom to see your signal. In cosmic terms, however, this colossal structure is ephemeral and will last only a few million years, gradually eaten away from above. The stars born in the original cluster and those made in the various pillars will then mingle before slowly dispersing into the Milky Way.

Address Second Galactic Quadrant, The Milky Way (21h 36m 50.1s, +57° 30' 57") | Getting there Head towards the constellation of Cepheus, the legendary king of Aethiopia, and travel for 2,400 years at the speed of light. | Tip This part of the Milky Way is littered with interesting nebulae, such as the famous North America Nebula, which looks remarkably like its namesake continent (20h 59m 26.7s, +44° 31' 19").

54 Galactic Center
Into the heart of the beast

To reach the center of our Milky Way galaxy, you'll need to travel towards the darkness. From Earth, our view of it is obscured by great clouds of gas and dust in the spiral arms that lie between here and there. But once you've penetrated those, there is a glorious chaos to behold.

As you might expect, the Galactic Center is crowded with stars. In the innermost few light years, they're packed 50 million times more closely than in the solar neighborhood. There are many old red giants and some white dwarfs, embers of even older stars. Viewing the region with your goggles set to infrared, you'll see young massive stars, some in clusters embedded in dense clouds, indicating active star formation. One of the dense clouds, Sagittarius B2, contains vast quantities and varieties of molecules, including ethanol, methanol, and a fruity ester, ethyl formate. Combine them, and you'll notice that the cloud smells of raspberry rum.

Add in radio waves, and the picture becomes even more complex, with delicate tracery around the star-forming regions, vast bubbles left over from recent supernovae, and many strange arcs and filaments crossing the galactic plane. These structures contain high-energy electrons spinning around strong magnetic fields, although their origins remain mysterious.

At the very center there's a bright radio source called Sgr A* found in 1933 by Karl Jansky (1905–1950), which also occasionally flares in x-rays. Watch carefully for a few years, and you'll see stars in the region moving on quick, tight curves as they fly close to the radio source and away again. From the shape and speed of those curves, you'll be able to deduce that they're orbiting a compact object with four million times the mass of our Sun: a supermassive black hole. Andrea Ghez and Reinhard Genzel shared the 2020 Nobel Prize in Physics for their two teams' work over decades leading to this discovery.

Address The Center of the Milky Way (17h 45m 40.0s, -29° 00' 28") | **Getting there** Head towards the constellation of Sagittarius, the archer, and continue for 27,000 years at the speed of light. | **Tip** A fifth of the way through your journey, you'll pass through a ring of star-forming regions called Sharpless 16 to 20. Because of their proximity to the line-of-sight to the Galactic Center, they're little studied, so take some notes and share your observations (17h 46m 22.5s, -29° 19' 52").

55 — HD209458b
The first known transiting exoplanet

Why bother taking a trip to see HD209458? After all, it's just one of many stars in the Milky Way that's broadly similar to our Sun, albeit a little bigger, younger, hotter, and more luminous. But if you have extremely sharp eyesight, you might notice something odd. Every three and a half days, like clockwork, the star appears to dim slightly, losing about two percent of its brightness. After three hours, it returns to its original intensity.

What's causing these regular dips in the brightness of HD209458? The answer lies not with the star, but rather a planet called HD209458b that orbits around it. For a traveler from Earth, the system is oriented edge-on, so that once every orbit, the planet crosses in front of the star for those three hours, temporarily blocking a little of the starlight. Astronomers call this passage a "transit."

In 1999, teams led by David Charbonneau and Gregory Henry saw these dips in the brightness of HD209458, opening a whole new way to discover and study planets orbiting other stars, so-called "exoplanets." For example, in a transiting system, if you know the radius of the star, then you can work out the radius of the exoplanet from how much the brightness drops. If you know its mass from other measurements, now you can calculate its density. And if the depth of the dip changes with wavelength, that'll indicate that the exoplanet has an atmosphere, and you'll be able to figure out which kinds of gases are in it. Those are very difficult measurements to make, though, and astronomers remain uncertain whether HD209458b's atmosphere holds any interesting molecules such as water.

What is sure is that you'll find a gas giant planet roughly 70 percent the mass of Jupiter. Since it orbits just 7 million km (4.3 million mi) from the star, it's heated to a toasty 750°C (1,380°F). Probably best not to plan for any camping on HD209458b.

Address Outer reaches of the Solar Neighborhood, The Milky Way (22h 03m 10.8s, +18° 53' 04") | **Getting there** Head towards the constellation of Pegasus, the mythical winged horse, and travel for 157 years at the speed of light. | **Tip** IK Pegasi is the nearest binary system to the Sun containing a star that might explode as a supernova. Fortunately, that probably won't happen for two billion years (21h 26m 26.7s, +19° 22' 32").

56 Herbig-Haro 212
A cosmic double lightsaber

To the northeast of the Horsehead, far from the vigorous, star-forming nurseries of the Orion and Flame nebulae, you'll find an apparently quiet region filled with clouds of gas and dust, cold and dark. But first impressions can be deceptive. Tune your goggles to far-infrared wavelengths to see a sparse field of very young protostars about 100,000 years old, betrayed only by a faint glow as they gently warm the material shrouding them. These are the next generation.

A giant disk rotates around one of those protostars, slowly pulling in material and directing it towards the growing central object. But magnetic fields trapped and twisted by the rotation divert and focus some of the material into a pair of jets that blast out from the north and south poles of the protostar. Moving at speeds of around 400,000 km/h (250,000 mph), the jets crash into surrounding material – the collision shocks and heats molecular hydrogen gas, making it shine in the near-infrared. So switch to those wavelengths, and you'll see your destination in all its glory: Herbig-Haro 212 (HH212), found in 1993 using the NASA Infrared Telescope Facility on Maunakea, Hawai'i.

Like Darth Maul's double red lightsaber, the HH212 jet extends symmetrically for more than two light years on either side of the protostar, with mirrored pairs of knots and bow shocks. Watch for long enough, and you'll see them expanding outwards. Conversely, trace those motions backwards in time and find that they're probably connected to explosive events thousands of years ago, as the young protostar burped after swallowing too much material at once.

But as spectacular and dynamic as the jets are, this is just a temporary phase. Return to HH212 in a million years' time, and the shrouding dust and gas will have been blown away by the jets and associated outflows, revealing a new star perhaps already circled by young planets.

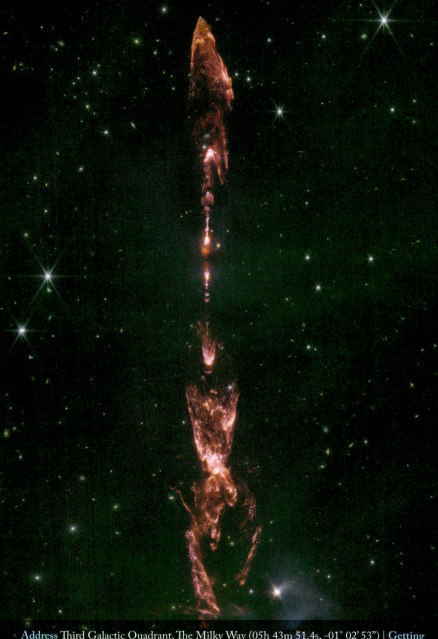

Address Third Galactic Quadrant, The Milky Way (05h 43m 51.4s, -01° 02' 53") | **Getting there** Head towards the constellation of Orion, the hunter, and travel for 1,300 years at the speed of light. | **Tip** Nearby lies Barnard's Loop , a giant, semi-circular arc of interstellar dust blown by a supernova two million years ago and illuminated by ultraviolet light from hot stars in the region (center 05h 30m, -04°).

57 Horsehead Nebula
A dusty equine near the hunter's belt

If you want to see Da Vinci's *Mona Lisa* in Paris or Van Gogh's *Starry Night* in New York, you'll have to brave the crowds – those artworks are as popular as they are unique. In space, the Horsehead Nebula is similarly popular. Gloriously resembling one of Earth's most elegant species, it's one of the most frequently photographed celestial objects. It's found within Orion, itself one of the most famous constellations in the sky.

The Horsehead, also known as Barnard 33, is a small, dark cloud of gas and dust discovered in 1888 by Scottish astronomer Williamina Fleming (1857–1911) while she was working as a "computer" at the Harvard College Observatory. It is seen silhouetted against the bright purple-blue nebulosity of IC 434, a region of hot plasma ionized by the nearby massive star Sigma Orionis. The surrounding molecular cloud has been sculpted by the wind and ultraviolet light of another star, Epsilon Orionis or Anilam, the central star in Orion's belt. A local enhancement in the density of the cloud has left a protruding, red-brown pillar that, some might argue, looks more like a seahorse than a terrestrial one.

Nevertheless, as you get closer, the horse begins to reveal an inner life. Near the top of its brow, you'll see a yellow star accompanied by a small, bright nebulosity. This is a young star that has been forming quietly within the dark confines of Barnard 33, but which has now been exposed as the radiation from the nearby massive stars further erodes the pillar.

Examining the pillar with your goggles set to infrared reveals a handful or two of other young stars forming inside, and radio frequencies show a number of dense gas clumps where more stars may form in the future. If you return a few million years from now, the nearby massive stars will have won, the Horsehead will have been washed away, and a small group of young stars is likely all that will remain.

Address Third Galactic Quadrant, The Milky Way (05h 41m 00.1s, -02° 27' 30") | **Getting there** Head towards the constellation of Orion, the hunter, and travel for 1,400 years at the speed of light. | **Tip** NGC2023 is a small cluster of stars near the Horsehead. The brightest, HD37903, is lighting up the surrounding cavity of gas and dust in delicate shades of blue, yellow, and red (05h 41m 38.4s, -02° 15' 32").

58 __ HR8799

An extrasolar orrery

Although ever advancing technology allows us to capture spectacular space images that were unimaginable just a few years ago, some places are so small or distant that we're lucky to get even low-quality images of them. But challenges like these don't mean that those places aren't worth visiting. The star HR8799, for one, is definitely somewhere you should see.

The star itself is fairly ordinary, about 45 percent more massive than our Sun and five times more luminous. Block its light, however, and something remarkable is revealed: a system of four planets. Christian Marois and collaborators discovered them between 2008 and 2010, and observations since then have shown them in motion, orbiting around HR8799.

What will you find when you get there? All four planets are gas giants a little larger than Jupiter and with six to nine times its mass. They span a range of distances from the star, the innermost one slightly closer than Uranus is to the Sun and the outermost more than four times farther out. You'd expect them to be cold, but surprisingly, all four are at temperatures between about 650°C and 900°C (1,200–1,650°F). Why is that? Because HR8799 is probably between 25 and 60 million years old, much younger than our Sun, and its planets still retain much of the warmth of their formation out of gas in the disk circling the star. Indeed, if you look carefully with your goggles set to infrared, you'll be able to see emission from debris left over from that original disk, still orbiting the star.

The planets' atmospheres are rich in molecules, including water vapor, carbon monoxide, ammonia, methane, and acetylene, and they're completely shrouded by clouds made of gaseous silicates, iron, and sodium sulfide. They are such strange, inhospitable worlds, and there's still much to learn and understand about them. The closer you can get, the more detailed a picture you'll build up.

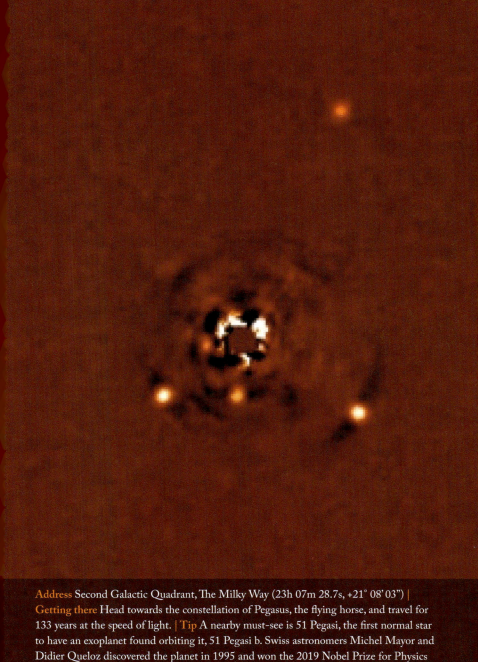

Address Second Galactic Quadrant, The Milky Way (23h 07m 28.7s, +21° 08' 03") | **Getting there** Head towards the constellation of Pegasus, the flying horse, and travel for 133 years at the speed of light. | **Tip** A nearby must-see is 51 Pegasi, the first normal star to have an exoplanet found orbiting it, 51 Pegasi b. Swiss astronomers Michel Mayor and Didier Queloz discovered the planet in 1995 and won the 2019 Nobel Prize for Physics (22h 57m 28.0s, +20° 46' 08").

59 Lagoon Nebula

Cloudy, with a chance of twisters

En route to the heart of our Milky Way galaxy, you'll pass through the darkness of the Great Rift, the giant gas and dust clouds in the spiral arms between Earth and the Galactic Center. These clouds are also the birthplaces of new stars. One of the largest star-forming regions you'll encounter on your way is the Lagoon Nebula. While the name suggests that it might be a tranquil place to pause on your journey, it's anything but.

The Lagoon Nebula is just visible to the naked eye and was noted around 1654 by Giovanni Battista Hodierna (1597–1660). It's also in Messier's catalog as M8. An HII region, a complex of hydrogen gas ionized by hot, young stars, it's much larger than the Orion Nebula. As you approach, you'll get the sense of a giant cavity having been excavated in a cloud with a ragged edge of dark pillars and globules around it, and clusters of stars around a million years old within. Your eye will be drawn to the brightest part of the nebula, where you'll see a tangle of red-brown dust and multi-colored gas known as the Hourglass. It is being illuminated, sculpted, and eroded by the violent winds and ultraviolet light of a hot, blue star at its center called Herschel 36, seen peeking out through a gap in the dust. In fact, Herschel 36 is likely to be a multiple system of at least three massive stars, similar to the Trapezium in Orion, surrounded by a very compact cluster of lower-mass ones.

There are many other signs of the hectic star-formation activity in the center of the Lagoon Nebula, including protoplanetary disks being blasted by the hot stars, and jets and outflows from other young stars. To add to the danger of stopping off here, it has been suggested that some of the huge columns of gas and dust around the Hourglass might be spinning, colossal versions of terrestrial tornados. Most astronomers believe that's quite implausible, but you've been warned.

Address First Galactic Quadrant, The Milky Way (18h 03m 40.3s, -24° 22' 43") | **Getting there** Head towards the constellation of Sagittarius, the archer, and travel for 4,200 years at the speed of light. | **Tip** The smaller Trifid Nebula is located close by and may be part of the same, large star-formation complex as the Lagoon. It's named for its three-lobed form broken by dust lanes, not the carnivorous plants of the 1951 novel *The Day of the Triffids* by John Wyndham (1903–1969) (18h 02m 23.5s, -23° 01' 52").

60 LL Pegasi
Two stars locked in a dusty death spiral

As you approach LL Pegasi, you won't be able to make out the heart of the region, which is hidden in a cloud of reddish-brown dust, *Peanuts'* Pigpen-style. But you will see the enormous pinwheel nebula surrounding it. Strangely, this dusty spiral is not illuminated by whatever is hidden in the core, but by the faint glow of our Milky Way galaxy.

Now look into the core with your goggles tuned to infrared. You'll find an old, pulsating, red giant star, far larger than the Sun, and with an atmosphere much richer in carbon than oxygen. This is the opposite of the normal state of affairs, so while some of the carbon reacts with oxygen to create carbon monoxide, the rest is free to form other carbon compounds, a kind of "soot." This material colors the star's atmosphere a brilliant ruby red before being ejected in its wind.

Over time, up to half the total mass of such a "carbon star" can be lost this way. In LL Pegasi, also known as AFGL 3068, it's flowing out in a sequence of beautiful spiral shells, a shape that matches calculations made by Archimedes (c. 287–212 BCE), also appearing in watch-springs, sunflower heads, and galaxy arms. By measuring the speed of the material moving away from LL Pegasi and the spacing of the shells, astronomers have found that each is separated by about 800 years. What's causing this regular structure? Look very closely into the core again: there's a second star. Its orbital period around the carbon star closely matches the spacing of the shells, and so the idea is that as its dust flows away, the gravity of its companion molds it into the expanding, three-dimensional, spiral sculpture you see.

But plan to visit LL Pegasi soon – this is a short-lived phase in a star's life, lasting only a few thousand years. And take a change of clothing because nothing stays clean after being sprayed with fine carbon soot moving at 53,000 km/h (33,000 mph).

Address Second Galactic Quadrant, The Milky Way (23h 19m 12.6s, +17° 11' 33"). | **Getting there** Head towards the constellation of Pegasus, the winged horse, and travel for 4,200 years at the speed of light. | **Tip** Alpha Pegasi, or Markab, is close on the sky to LL Pegasi but much nearer to Earth. Hot and blue, this star has run out of hydrogen fuel, making it swell in size (23h 04m 45.7s, +15° 12' 19").

61 The Milky Way
Seeing the forest for the trees

Imagine you're in Spain on a clear, moonless night, far from any lights. You look up and see a few thousand stars distributed randomly all over the sky, some bright, many faint. Repeat six months later and there will be many new stars. Do the same in New Zealand, and there will be yet more.

From this, you might deduce that Earth is surrounded by a simple sphere of stars fading into the distance. However, the symmetry is broken by a pale band of light across the sky, occasionally interrupted by dark clouds. In many cultures, this band is associated with milk, while others believe it to be a river or path for fish, birds, or souls. It is a clue that perhaps the Universe around us is curiouser than you first thought.

Now imagine flying perpendicular to this band until you're a million light years away. Look down, and all is revealed. You see a flattened disk 100,000 light years across made of hundreds of billions of stars. At the center is a bar-shaped bulge of older, yellow ones. From it, gas, dust, and younger stars trail in spiral arms. The Solar System is in one of those arms, halfway to the edge of the disk. If you were very patient, you'd see stars slowly orbiting the center, the Sun taking 230 million years to do so. Behold the Milky Way, your home galaxy.

Now, imagine you're back on Earth. While we *are* surrounded by a more or less uniform ball of stars locally, the Milky Way's structure becomes important on larger scales. The hazy band of light is the galaxy seen edge-on, brighter as you look towards its center and fainter in the opposite direction. Dusty clouds in nearby spiral arms partially block your view of more distant ones, and companion galaxies, the Nubecula Major and Minor, appear near the band. Finally, imagine how we gained all of this knowledge with telescopes, satellites, and our own brains – without actually ever leaving home.

Address The Local Group, Virgo Supercluster, Laniakea Supercluster, Deep Space | **Getting there** To see the Milky Way from above, head towards the constellation of Coma Berenices and travel for a million years at the speed of light. To see it from below, head in the direction of Sculptor instead. | **Tip** After just 29,000 years of your journey to the South Galactic Pole in Sculptor, you'll pass the globular cluster NGC288. Its core is surrounded by a partial ring of brighter stars (00h 52m 45.2s, -26° 34' 57").

62 NGC1999

When is a hole not a hole but actually is a hole?

If you look up at the misty starlight of the Milky Way on a clear, moonless night, you'll see a long dark band along it called the Great Rift. The Inca named parts of it in the same way as constellations, including the llama, the snake, and the partridge. Well into the 20th century, the Great Rift was thought to be holes and gaps in the distribution of stars until astronomers E. E. Barnard (1857–1923) and Max Wolf (1863–1932) recognized that it's due to nearby clouds of gas and dust obscuring the stars beyond. This discovery of dark nebulae and similar objects, such as Bok globules, across the Milky Way and often associated with regions of star formation started a revolution in our understanding of how stars are born.

But sometimes, received wisdom gets turned upside down, as you'll discover when you visit NGC1999. Just south of the Orion Nebula, NGC1999 is one of many pockets of star formation in the molecular clouds in Orion's sword and belt. With your goggles set to infrared, you'll see young protostars embedded in their natal gas and dust, ejecting high-speed jets and outflows of gas from their protoplanetary disks. The heart of NGC1999 is a blue nebulosity reflecting the light of a bright, variable star V 380 Ori, ringed by dusty brown clouds. And at its center is an irregular, dark blob that must surely be one of the densest dark clouds known.

Except it's not: it really is a hole, or an absence, called the Cosmic Keyhole. Observations at infrared and millimeter wavelengths made in 2009 confirmed that there's a much lower density of dust and gas there than in the surroundings. That's not to imply that it's some kind of weird alien portal, but exactly how such a sharp-edged hole was made remains a mystery. One theory is that a jet from V 380 Ori punched through the local cloud so recently that there hasn't been enough time for the hole to heal. You won't have any problem flying through the hole though – it's a trillion km (620 billion mi) wide.

Address NGC1999, Third Galactic Quadrant, The Milky Way (05h 36m 24.5s, -06° 42' 56") | **Getting there** Head towards the constellation of Orion, the hunter, for around 1,300 years at the speed of light. | **Tip** Near NGC1999, you'll find HH1 & 2, a pair of bright knots of ionized gas ejected by a young protostar. They were the first objects in the important Herbig–Haro class, named for their discoverers George Herbig (1920–2013) and Guillermo Haro (1913–1988) (05h 36m 22.9s, -06° 46' 10").

63 The OMC-1 Explosion
Cosmic shrapnel

Many people enjoy potholing and spelunking, exploring dark underground caves perhaps in the hope of making stunning discoveries like the Cueva de los Cristales in Mexico, filled with huge selenite columns that grew slowly over more than half a million years. There is a parallel with astronomers who study the skies at infrared wavelengths, peering through the dust and gas that shroud so many parts of the Milky Way to discover new stars. One such location is a dark, dense cloud in Orion, where an exciting array of much larger "crystal columns" was created by an altogether more dramatic event.

The Orion Molecular Cloud Number 1 or OMC-1 lies behind the Orion Nebula and is filled with cold dust and gas. In some denser parts, the cloud has contracted under the force of gravity to make protostars, and if you switch to infrared vision, you'll see their glow. One key, small cluster lurks in the cloud just northwest of the Trapezium stars, signposted by a spectacular spray of "fingers" extending up to a light year from its heart. Along most of their turbulent length, the fingers glow red due to emission from molecular hydrogen gas shocked and heated by collisions. Their tips often turn green thanks to temperatures high enough to light up gaseous iron.

You'll see that the fingers are racing outwards at speeds of up to 720,000 km/h (450,000 mph). In turn, you can trace them backwards to realize that something remarkable must have happened around 500 years ago in the core of the cluster. It's thought that two young protostars collided, releasing a huge explosion of energy. A shockwave expanded outwards, causing the surrounding cloud to fragment into "bullets" of dense material that are still flying away from the scene, trailing wakes of hot gas. Within a million years or so, the dust around the protostellar cluster should have cleared away, bringing it out of the dark cave, so to speak.

Address Third Galactic Quadrant, The Milky Way (05h 35m 13.4s, -05° 21' 44") | **Getting there** Head towards the constellation of Orion, the hunter, and travel for 1,270 years at the speed of light. | **Tip** OMC-2 is another dense molecular cloud to the north of the Orion Nebula with a cluster of young protostars believed to be typical of the environment in which our Sun and Solar System were born (05h 35m 26.7s, -05° 10' 00").

64 Omega Centauri
King of the globular clusters

If you're inspired by the uncluttered forms and minimal ornamentation of Bauhaus architecture, you should plan a trip to see a globular cluster. They are among the simplest stellar systems known. Devoid of gas and dust, they comprise a ball of stars, from tens of thousands to millions, held together by their mutual gravity. Typically, they're found distributed in a spherical halo around a bigger galaxy – for example, the Milky Way has at least 150 globular clusters.

The brightest and largest of them is Omega Centauri, which has been known since antiquity. Ptolemy (c. 100–170 CE) included it in his famous catalogue *The Almagest* as just a single star, even though it's fuzzy and almost as large as the full Moon when seen from a dark site. Edmond Halley (1656–1742) of the eponymous comet viewed it through a telescope from Saint Helena in 1677, noting that it was "extended," and James Dunlop (1793–1848) later described it as a "beautiful globe of stars."

When you arrive at Omega Centauri, you'll see around 10 million stars. Continue to the core, and your sky will be filled with them, seemingly within touching distance. The density is so high there that stars are typically just 0.1 light years apart, far closer than the 4.2 light years between the Sun and our nearest stellar neighbor, Proxima Centauri.

For a long time, astronomers believed that stars in normal globular clusters were almost as old as the Universe, having formed in one go out of a single cloud of gas and dust. But modern studies of the stars' compositions shows that they're more likely made episodically over time. Omega Centauri is special though – it's probably the core of a dwarf galaxy that had a close encounter with the Milky Way and was then disrupted and captured. The fast-moving Kapteyn's Star, which is close to the Sun today, may have been ejected at high speed from Omega Centauri during this event.

Address Orbiting the Milky Way (13h 26m 47.2s, -47° 28' 46") | **Getting there** Head towards the constellation of Centaurus, the centaur, and travel for 15,800 years at the speed of light. | **Tip** In the heart of Omega Centauri, rapidly-moving stars are the telltale sign of the presence of a black hole with at least 8,200 times the mass of our Sun (13h 26m 47.2s, -47° 28' 46").

MILKY WAY

65 — Orion Nebula
Massive star formation below the belt

Orion is the most easily recognized constellation in the night sky, and it's fitting that it's also home to one of the most iconic star-forming regions, the Orion Nebula. With the naked eye, you can just make out a faint fuzziness around some stars in the hunter's sword dangling below the three stars of his belt. But the full glory is revealed by magnifying the view – or getting closer.

The nebula is known as an H$_{II}$ region. You'll see it glowing in reds, yellows, greens, and blues thanks to hydrogen, helium, oxygen, nitrogen, carbon, neon, and other ionized gases. Dust is also important, as it reflects light in some places, while in others, it lies in brown tendrils across the face of the nebula like a half-drawn curtain. There's a visceral sense of a cavity carved out of a huge wall, and indeed that wall is a giant molecular cloud behind the visible nebula.

At the nebula's heart, you'll spot the Trapezium, a tight group of four stars between 7 and 45 times the mass of the Sun that is lighting up the nebula. Most of these massive stars are in binaries or higher-order multiple systems. Switch to infrared vision to cut through the haze of gas and dust, and you'll see thousands of smaller stars crowded around them – this is the Trapezium Cluster, only about a million years old. There are also many brown dwarfs, objects with less than seven percent of the mass of our Sun that are unable to fuse hydrogen in their cores. Some are just a few times more massive than Jupiter, and exactly how they formed remains a mystery.

Get closer, and you'll find that many of the stars and brown dwarfs are surrounded by dense disks of gas and dust. It's likely that planets are forming in the disks, but in many cases, the disks are also being eroded by the ultraviolet light and winds of the massive stars in a race against time. There's plenty of other action going on in the Orion Nebula, including jets of gas expanding from young stars, making it an exciting destination to explore.

Address Third Galactic Quadrant, The Milky Way (05h 35m 17.0s, -05° 24' 47") | **Getting there** Head towards the constellation of Orion, the hunter, and travel for 1,270 years at the speed of light. | **Tip** The Running Man Nebula (Sh 2-279) lies to the north of the Orion Nebula in the same star-forming complex. It has fewer hot massive stars, but it boasts a beautiful, red and blue H$_{II}$ region and reflection nebula (05h 35m 16.2s, -04° 48' 27").

MILKY WAY

66 Pillars of Creation
Towering columns of gas, dust, and young stars

If you're heading towards the center of the Milky Way, you'll traverse our galaxy's spiral arms. Sprinkled across them are regions where new stars are being born from clouds of dark dust and brightly-glowing gas, and many are worth visiting. One of the most famous of these regions, the Eagle Nebula, was discovered by Jean-Philippe Loys de Cheseaux (1718–1751) and cataloged as M16 by Charles Messier (1730–1817).

Travel to the Eagle, and you'll see a dense cluster of stars called NGC6611 that formed only a few million years ago. The largest is around 80 times the mass of the Sun and a million times brighter. Intense ultraviolet light from it and its massive siblings ionizes and energizes tenuous hydrogen gas to illuminate a bright nebula resembling a raptor with outspread wings. The light and wind from these stars is also sculpting and eroding dark, dense columns of gas and dust, each several light years long, seen silhouetted against the Eagle's glow. In 1995, the Hubble Space Telescope took pictures showing a trio of columns in brown and yellow against a blue-green backdrop, like a giant, undersea kelp forest. Astronomers found hints that the columns may contain newly-born stars, perhaps formed as the gas was compressed from the outside, leading to them being nicknamed the "Pillars of Creation."

Tune your goggles to infrared, like the NASA/ESA/CSA James Webb Space Telescope, and more is revealed. Peering through the dust that shrouds those baby stars, you'll see some are accompanied by flame-red jets of outflowing material, a sign of their youth. The craggy pillars themselves become more transparent and reveal internal structures, also taking on an orange glow as complex hydrocarbon molecules on their surfaces are lit up by the ultraviolet light. And beyond the Eagle Nebula, you'll find a myriad of background stars, marking the way on your journey ever deeper into the galaxy.

Address First Galactic Quadrant, Sagittarius Arm, The Milky Way (18h 18m 55.2s, -13° 51' 15") | **Getting there** Head towards the constellation of Serpens Cauda, the serpent's tail, and travel for 5,700 years towards the inner galaxy at the speed of light. | **Tip** NGC6604 is an older star-forming region on the way to the Eagle Nebula and hints at how the latter may look in a few million years' time after the pillars have been destroyed (18h 18m 05.8s, -12° 14' 35").

MILKY WAY

67 The Pleiades
The star cluster with a thousand names

Subaru, al-Thurayya, Nyarluwarri, Tŵr Tewdws, Kṛttikā, Πλειάδες, Ngauponi, Messier 45 – the list of names for this iconic, open star cluster spans continents, millennia, and almost all of human culture. It even features prominently on the Nebra sky disk, a 3,700-year-old, bronze-and-gold guide to the heavens dug up by two treasure-hunting looters atop a hill in eastern Germany in 1999, the oldest known representation of actual astronomical phenomena.

But you won't need to use the sky disk to navigate towards the Pleiades, one of the most obvious clusters of young stars in the sky. As you head towards it, you'll first be taken by the diaphanous, blue nebula that swirls around the famous Seven Sisters: Maia, Electra, Taygete, Alcyone, Celaeno, Sterope, and Merope, daughters of Atlas and Pleione. They are all hot, blue stars, and their light reflects off dust in their vicinity. Farther away, the dust turns to a mix of reds and browns. Astronomers once believed that this dust was left over from the formation of the Pleiades 125 million years ago, but it's now thought that the cluster is simply drifting through a particularly dusty part of our Milky Way galaxy.

As you get closer to the Seven Sisters, many fainter siblings will come into view – there are more than 1,000 lower-mass stars and sub-stellar brown dwarfs in total. It's not a large cluster, but due to its proximity to Earth, it has been a prominent feature of the night sky throughout human history.

The Pleiades won't last forever though. The bright sisters are already approaching the end of their lives on the main sequence. They will soon become red giants and then white dwarfs. Their smaller companions will continue to burn more modestly for billions of years to come, but long before that, the cluster will slowly expand and dissolve into the Milky Way, joining the hundreds of billions of other stars.

Address Second Galactic Quadrant, The Orion Arm, The Milky Way (03h 47m 24.0s, +24° 07' 00") | **Getting there** Head towards the constellation of Taurus, the bull, and go for 444 years at the speed of light. | **Tip** The Hyades is another famous, open star cluster in Taurus. It is older than the Pleiades and closer to Earth, spanning a much larger area of the sky. Its brightest star is the red giant Aldebaran (04h 35m 55.0s, +16° 30' 33").

68 Polaris

The North Star ... for now at least

Ideally, you'll begin this journey from the North Pole at night. Look directly up to the zenith – that's the North Celestial Pole. There's a bright star close to it called Alpha Ursae Minoris. It plays a special role in human culture because of its location in the sky, giving it the more common name of Polaris. Set the controls to rise vertically, and you're on your way.

As Earth spins on its axis, the Sun, planets, and stars appear to move across the sky. They make circles every 24 hours around the North or South Celestial Pole, depending on which hemisphere you're in. While there's no star near the South Celestial Pole, there is at the North one, and that's Polaris. It's less than one degree from the pole, about 1.5 times the angular diameter of the Moon, but only temporarily. Earth's rotation axis precesses and makes a large circle on the sky every 26,000 years. This means that Polaris is only near the pole for part of that period, and other stars take their turns during the cycle. For example, Alpha Draconis, also known as Thuban, was very close when Egyptian pharaoh Djoser ordered the construction of the first pyramids some 4,700 years ago.

As you get closer to Polaris, you'll realize that it is, in fact, a triple system. The brightest of the three is a yellow supergiant about five times the mass of our Sun. It is also a special kind of star called a Cepheid variable and pulsates in size and temperature every 3.9 days. Strangely, it seems to be much younger than either of its companions, something that remains baffling to this day.

As you look at the wider region surrounding the triple system, you'll notice very faint clouds of gas and dust glowing. They are not lit up by Polaris, however, and lie far beyond three stars and well outside the main body of our galaxy. Astronomers call these clouds an "integrated flux nebula" because it's illuminated by the general glow of the Milky Way, not any specific star.

Address Second Galactic Quadrant, The Milky Way (02h 31m 49.2s, +89° 15' 51") | **Getting there** Head towards the constellation of Ursa Minor, the little bear, and continue for 448 years at the speed of light. | **Tip** NGC188 also lies near the North Celestial Pole and is one of the most ancient open star clusters in the Milky Way, having survived for several billion years (00h 47m 36.3s, +85° 15' 19").

69   R Aquarii
A story of symbiosis between little and large

Binary and multiple stars account for about one-third of all stellar systems in our galaxy. If the two members of a binary have different masses, they will evolve at different speeds, which can lead to some spectacular interactions between them, as you'll see when you reach R Aquarii, a symbiotic binary system that sits inside the colorful nebula Cederblad 211.

In this system, one of the stars is a tiny but hot, white dwarf that has sloughed off its outer envelope and stopped nuclear fusion. The other star is a cool, red giant, a Mira variable, pulsating and changing brightness by a factor of 750 every 13 months. You'll see that the two stars are dancing around each other in a highly elliptical orbit. When the stars come close to each other every 44 years, the gravity of the white dwarf distends the bloated red giant, pulling material off its surface and from its dense stellar wind. That material spirals towards the white dwarf through an orbiting accretion disk. Much of the material ends up on the white dwarf, but some is diverted outwards into polar jets moving away from the white dwarf at around 2.6 million km/h (1.6 million mph).

As the jets precess, the hot flowing gas in them is twisted to form the vertically-oriented, S-shaped structure you see. Farther out, there are rings and shells of gas, suggesting that other violent processes may be at work. This takes us back to the white dwarf. Over time, you'll see hydrogen gas pulled from the red giant building up on its surface. It gets heated up by the white dwarf, and once it reaches a critical temperature, runaway nuclear fusion can occur, creating a sudden flash of energy. It's thought that R Aquarii becomes such an explosive "nova" every few hundred years. Korean astronomers recorded a pair of outbursts in 1073 CE and 1074 CE at the position of R Aquarii, and these may be linked to some of the nebular structures we see today.

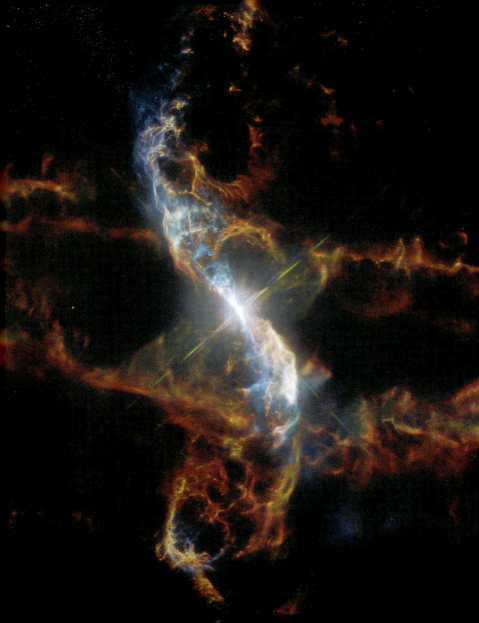

Address First Galactic Quadrant, The Milky Way (23h 43m 49.5s, -15° 17' 04") | **Getting there** Head towards the constellation of Aquarius, the water carrier. The distance is uncertain, and so it may take anywhere between 600 and 850 years at the speed of light. | **Tip** On your way to R Aquarii, you'll pass two interesting multiple systems, Omega-1 Aquarii, a binary, and Omega-2 Aquarii, a triple. Both are part of the Palace Guard asterism in the Chinese Encampment mansion (23h 41m 15.0s, -14° 23' 00").

70 Ring Nebula
When is a planet not a planet?

When perusing lists of possible space destinations to visit, you might occasionally be puzzled by the nomenclature. One confusion that persists to this day is the name "planetary nebula," used for objects which have nothing to do with actual planets like Mars or Saturn. When first noticed in the 1700s, some were seen to be round and glow more or less uniformly, and they were described as looking like planets. Astronomers knew perfectly well that they weren't actual planets, as they didn't move through the background of stars, but the name has stuck. Charles Messier discovered many while sweeping the skies for comets, including one he gave the number 57 in his famous catalog, now better known as the Ring Nebula.

As you approach, you'll see that it has a central blue-green oval with bright concentric annuli in yellows and reds. These colors tell you that there are different ionized gases involved, including oxygen, hydrogen, and nitrogen, as well as complex organic molecules. At the center of the nebula, you'll spot a faint, white, dwarf star. It seems innocuous enough today, but only 2,000 years ago, it was a huge, cool giant star about to run out of nuclear fuel. As it did, its outer atmosphere swelled, and half of the star's mass blew outwards into a giant bubble. When the remaining core collapsed, it formed the hot white dwarf that now supplies the ultraviolet light ionizing the surrounding gas.

The bubble is actually hourglass-shaped with a torus of dense gas around the equator, but because you're seeing the Ring Nebula almost end on, it appears more or less round. As you get closer still, you'll spot thousands of small gas globules in the nebula being blasted by the central star. Farther out, there are tendrils and fainter arcs around the nebula which astronomers think may have been shaped by another star as it orbited the dying giant.

Address First Galactic Quadrant, The Milky Way (18h 53m 35.1s, +33° 01' 45") | Getting there Head towards the constellation of Lyra, the harp, and continue for 2,500 years at the speed of light. | Tip Adjacent to the Ring Nebula is another fuzzy object that you might also consider visiting. But don't be fooled – IC1296 is a spiral galaxy and lies 238 million light years further on (18h 53m 18.8s, +33° 04' 00").

71 RS Puppis
Rhythm of the stars

Long before you reach RS Puppis, you'll see it signposting the way, pulsing slowly like a cosmic lighthouse. You'll also know how much farther you have to travel by measuring its apparent brightness. This star is a yellow supergiant and a Cepheid variable, one of the brightest known in the Milky Way. These stars have played a key role in the way we determine distances in our vast Universe.

Aged just 20, the deaf amateur astronomer, John Goodricke (1764–1786) made observations from a top floor window of his parents' home immediately adjacent to the gothic cathedral in York. He discovered regular variations in the brightness of Beta Lyrae and Delta Cepheid, the latter being one of the first of a class of variable stars now called Cepheids. For this and other work, Goodricke was elected as a member of the Royal Society at 21 but died just four days later from pneumonia before hearing the news. Later, Henrietta Swan Leavitt (1868–1921) discovered that the pulsation period of a Cepheid is directly related to its intrinsic brightness: the brighter the star, the slower the pulsations. She realized that Cepheids are "standard candles": if you can measure the pulsation period and apparent brightness of one, you can calculate its distance. And because Cepheids are so naturally bright, they can be seen at great distances, making them one of the most important rungs on the "cosmic distance ladder."

As you near RS Puppis, you'll see that it changes in size and temperature over a 41.5 day period as helium gas in its atmosphere rhythmically contracts and heats, then expands and cools again. Over the cycle, it brightens by 250 percent when it's at its largest – more than 200 times wider than our Sun – before shrinking and dimming again. The star is also surrounded by a cloud of dust and gas perhaps left over from its birth and if you observe very carefully, you'll notice moving ripples or "light echoes", reflecting the regular changes in brightness of RS Puppis itself.

Address Third Galactic Quadrant, The Milky Way (08h 13m 04.2s, -34° 34' 43") | **Getting there** Head towards the constellation of Puppis, the stern of the great ship of Argo Navis, and travel for 5,700 years at the speed of light. | **Tip** ASASSN-21qj is a young star in Puppis that showed a dramatic fading in 2021, which astronomers suspect was due to the collision of two ice-giant planets. Well worth a visit to see the debris that temporarily blocked the starlight (08h 15m 23.3s, -38° 59' 23").

72 __ Serpens Nebula
A hissing nest of star formation

If you're like Indiana Jones and suffer from ophidiophobia, perhaps this isn't the trip for you. Not only will you be heading to a constellation whose name means "serpent's tail", but when you get there, you'll be surrounded by long structures like a nest of giant snakes.

In reality, there's nothing to be afraid of. This very busy region of the Serpens Nebula contains a cluster of young stars that are perhaps only 100,000 years old, lighting up the gas and dust out of which they're being made in blues, yellows, reds, and browns. Many of them are encircled by dense disks of material that's swirling around as it feeds the central protostar. Some of that material escapes being swallowed, however, and makes planets, while another fraction ends up being ejected in jets from the north and south poles. As the jets crash into surrounding clouds, molecular hydrogen gas in the jets and clouds heats up and glows. Tune your goggles to infrared to see the writhing snake-like jets glow red.

Near the center of the region, you'll see another young star surrounded by wispy, blue-white clouds. Curiously though, they're interrupted on either side of the star by a pair of narrow, dark triangles, tapering towards the center. If you watch for long enough, the dark triangles will dance around. They are shadows being cast by the central protostar and its disk, and as the lumpy disk rotates and lets more or less light from the protostar escape, the shapes of the shadows change.

Within a few million years, most of the gas and dust in this region will have been cleared away, either eaten by the young stars or blown away by their winds. Further in the future, the cluster will slowly fall apart and dissolve into the Milky Way, and there will be no sign any more of its furious early history as the stars were first being forged. So overcome any irrational fear of snakes you might have and visit now before it's too late.

Address First Galactic Quadrant, Serpens-Aquila Rift, The Milky Way (18h 29m 55.3s, +01° 14' 32") | **Getting there** Head towards the constellation of Serpens Cauda, the serpent's tail, and travel for 1,600 years towards the inner galaxy at the speed of light. | **Tip** Westerhout 40 is a nearby region where much bigger stars are being born, but it's hard to see because of the dust around it. Tune to infrared, radio, or x-ray wavelengths for the full experience (18h 31m 24.0s, -02° 04' 57").

73 — Sirius
Twinkle, twinkle, little star

Of all the stars in a planet's night sky, only one can be crowned the brightest. For Earth, that star is Sirius. It features in myths, legends, and astronomical records across human history and cultures, and as the brightest star in the constellation of Canis Major, one of Orion's hunting hounds, it's also known as the Dog Star.

With double the mass and 25 times the luminosity of the Sun, Sirius is also one of the 10 closest stars to our Solar System. This combination is what makes it so special to us. It also hides a secret: invisible to the naked eye is a companion, Sirius B, discovered only in 1862 by Alvan Graham Clark (1832–1897). Sirius B has the same mass as the Sun, but as the dying white dwarf ember of a larger red giant, it's now smaller than Earth and thus relatively faint.

As you approach, you'll see Sirius A and B slowly moving around each other on their 50-year orbit. Close up, you'll see that they're blue-white: they're both much hotter than the Sun, giving them this color. That might confuse you if you'd looked up at Sirius from Earth before you left on this trip. From there, Sirius seems to be a star of many hues, twinkling through intense blues, reds, yellows, and greens many times a second. But that's an illusion due to our turbulent atmosphere. Small changes in temperature lead to changes in the refractive index. As a result, the blue-white light splits into its rainbow components, yielding a flickering, technicolor display that can be captured with just a mobile phone camera.

Dig deeper into the historical record, though, and you'll find descriptions of Sirius as being red. Was that just the effect of the star twinkling, seen close to the horizon and reddened like a sunset, or was Sirius B still a red giant as recently as 2,000 years ago? Perhaps you'll find some clues during your visit to one of the most familiar stars in our sky.

Address Third Galactic Quadrant, The Milky Way (06h 45m 08.7s, -16° 42' 58") | **Getting there** Head towards the constellation of Canis Major, the greater dog, and travel for 8.6 years at the speed of light. Set your course carefully – Sirius is moving across the sky and slowly towards us. | **Tip** Once you arrive in the Sirius system, make a careful survey for any other companions – rumors have abounded for years that there may be a brown dwarf or even a planet lurking there.

74 Taurus-Auriga Clouds
A low-density neighborhood for young stars

One of the most accessible places to see young stars and planets being made today is in the nearby archipelago of gas and dust known as the Taurus-Auriga dark cloud complex. The complex is not making giant stars similar to those that light up regions like the Orion Nebula, but there are many young stars, perhaps just one to two million years old, with masses similar to that of our Sun or smaller. To visit them all, you'll need to navigate a vast, three-dimensional system of reefs and islands spanning more than 100 light years.

The complex of clouds can be traced across the sky at visible wavelengths, obscuring the light from more distant stars of the Milky Way. The clouds form a network of filaments, and dotted at points along them are signs of star birth – partly-hidden young stars lighting up their environment, some with planet-forming disks spinning around them and jets of gas spewing from their poles. When you tune your goggles to infrared, you'll be able to see through the dust and make a more complete census of the young population. Switch to even longer wavelengths, and the clouds themselves glow, thanks to dust warmed to just a few tens of degrees above absolute zero and gas molecules such as carbon monoxide.

The structure of the Taurus-Auriga clouds offers a clue to their origin. The filaments perhaps formed from larger clouds as the energy from nearby young stars compressed the cloud and broke it up. You can see material flowing along the filaments towards the places where the new stars are being made, while additional gas and dust are drawn into the filaments sideways thanks to magnetic fields. The Taurus-Auriga dark clouds are just one part of a much larger complex of star-forming regions known as the Radcliffe Wave, that runs along the local Orion-Cygnus arm of our spiral galaxy. It was discovered as recently as 2020, using data from ESA's Gaia mission.

Address Second Galactic Quadrant, The Milky Way (04h 30m 04s, +25° 30') | **Getting there** Head towards the constellation of Taurus, the bull, and travel for 450 years at the speed of light. | **Tip** A must-see is HL Tauri. Perhaps only 100,000 years old, this young star has a disk with a spectacular set of rings, and in the gaps between them, new planets are probably being built (04h 25m 05s, +26° 35' 52").

75 — Terzan 5

Digging for ancient galactic fossils

As with most galaxies, our Milky Way is surrounded by a halo of globular clusters. These are spherical balls of tens of thousands to several million stars compacted into a region tens to hundreds of light years across. They are free of gas and dust, and while it was once thought all of the stars were uniformly old, it's now known that globular clusters can contain several generations of stars. But, as you'll discover when you reach Terzan 5, this age spread can get quite extreme. Indeed, astronomers wonder whether it is a globular cluster at all or something altogether more exotic.

Even reaching Terzan 5 is quite a journey, as it lies towards the center of our galaxy, orbiting the bulge. Because of all the dust in that direction, it's quite hard to see and was only discovered in 1968 by Agop Terzan (1927–2020). You'll find that the stars at its core are incredibly close together, with a density perhaps 10 million times higher than around the Sun. Equally remarkably, Terzan 5 has two separate populations of stars mixed together, one around 4.5 billion years old, about the same age as the Sun, and the other 12 billion years old, closer to the age of the Universe. They also have different chemical signatures, indicating that they were made out of different mixtures of hydrogen, helium, and heavier elements. These characteristics all run counter to how astronomers think globular clusters are made.

Terzan 5 could be the result of two different clusters that merged. More likely, though, it's the remains of a dwarf galaxy that was disrupted and swallowed by the Milky Way, its core somehow remaining intact. If so, then it can give us clues about the way our galaxy was assembled out of smaller, primordial building blocks over many billions of years. It's almost a living fossil, helping us understand the evolution of the Milky Way, a kind of astronomical coelacanth.

Address First Galactic Quadrant, The Milky Way (17h 48m 04.9s, -24° 46' 45") | **Getting there** Head towards the constellation of Sagittarius, the archer, and travel for 18,800 years at the speed of light. | **Tip** On your way to Terzan 5, you'll pass close to the Silkworm Nebula, a very luminous giant star near the end of its life and shedding its outer envelope. The envelope will become a planetary nebula when the star shrinks again and heats up (17h 47m 13.5s, -24° 12' 51").

76 Vela Supernova Remnant
A cosmic memento mori

As you look at the gloriously colorful filaments and sheets of the Vela supernova remnant, consider that just 20,000 years ago, much of that glowing material was concentrated into a much smaller volume, namely a star. However, the mix of atoms that went into it 50–100 million years earlier is not the same as that which came out. Stars are alchemists, turning hydrogen and helium into many of the other elements in the Universe.

Stars are powered by nuclear fusion in their cores, where extreme pressures and temperatures turn hydrogen into helium, liberate energy, and resist the pull of gravity. In more massive stars, carbon, nitrogen, and oxygen act as catalysts, and even more complicated processes happen in stars with more than eight times the mass of the Sun, finally making elements as heavy as iron. Iron is a dead end for a star though, as it can't be fused into anything else without consuming energy, rather than liberating it. Unable to support itself against gravity, the star's time is over. The core collapses at almost 25 percent of the speed of light. As the central density rises to extreme values, it rebounds outwards less than a second later. This action creates a shockwave with extremely high temperatures, and, in an instant, many other elements are forged. As the star explodes, these elements are distributed back into space, along with other products of the star's lifetime, leaving just a fraction of the mass remaining as a black hole or neutron star.

Other processes, such as exploding white dwarfs, dying low-mass stars, and merging neutron stars, help populate the periodic table with all but the most extreme elements, which can only be made in a lab. So, as you continue to gaze at the beauty of the Vela supernova remnant, also consider the extreme conditions that made much of the material you're seeing, most of your spaceship, and about 90 percent of your mass.

Address Gum Nebula, Third Galactic Quadrant, The Milky Way (08h 31m 16.1s, -43° 37' 05") | **Getting there** Head towards the constellation of Vela, the sails of Argo Navis, and travel for approximately 1,000 years at the speed of light. | **Tip** Try to spot the neutron star that remained after the star that made the Vela supernova remnant exploded. More massive than our Sun but only 20 km (13 mi) in diameter, it is spinning 11 times per second, which makes it a pulsar (08h 35m 20.7s, -45° 10' 35").

77 — Westerlund 1

A giant lurking behind a dark veil

Living inside a flattened spiral galaxy means that our view of it is distorted, not least by nearby dark clouds of gas and dust that pepper the bright river of the Milky Way, blocking our view of things lying at greater distances. This situation accounts for the relative obscurity of Westerlund 1. Partly hidden behind clouds in the Scutum-Crux spiral arm, it was only discovered in 1961 by Bengt Westerlund (1921–2008) but is now thought to be the most massive young, super star cluster in the entire galaxy.

One way to get a clearer view en route to Westerlund 1 is to tune your goggles to infrared. Dust becomes more transparent at those wavelengths, allowing you to see the cluster in its full glory. Just a few million years old, it holds tens of thousands of stars packed into a small volume, including an impressive collection of very bright, very massive ones. Some of these stars are already near the end of their lives and have evolved into blue Wolf-Rayet stars, yellow hypergiants, and red supergiants.

Among them is Westerlund 1-26, one of the largest and brightest stars known. If placed in our Solar System, it would reach out the distance of Jupiter with a radius more than 1,000 times that of our Sun – and outshine it by a factor of more than 200,000.

Westerlund 1-26 is surrounded by a small nebula of ionized hydrogen, while several other massive stars have comet-like "tails" of dense material streaming away from them, blown in the cumulative powerful wind of the cluster. Look more widely in the cluster to see tattered, red remnants of molecular clouds being blasted by the immense radiation of the massive stars.

One puzzle worth investigating while you're here is why there's little sign of massive stars having already ended their lives in supernova explosions. There should have been 100 or so in the past million years, but so far, firm evidence for just one has been found.

Address Fourth Galactic Quadrant, Sagittarius-Carina Arm, The Milky Way (16h 47m 03.8s, -45° 50' 47") | **Getting there** Head towards the constellation of Ara, the altar, and continue for 13,800 years at the speed of light. | **Tip** On your way, you'll see HD148937, a pair of massive stars. They're surrounded by a beautiful, double-ringed nebula called NGC6164, also known as the Dragon's Egg (16h 33m 52.5s, -48° 06' 41").

MILKY WAY

78 — Zeta Ophiuchi
Big star in a hurry

To visit the massive star Zeta Ophiuchi, you'll have to calculate your trajectory with care. It is orbiting the Milky Way, but it also has an extra "peculiar velocity" of at least 72,000 km/h (45,000 mph) as it speeds through the dust clouds of its namesake constellation, creating a spectacular bow shock.

Massive stars are often found in multiple systems and tight groups at the center of clusters of young stars, illuminating the gas and dust from which they were born. As the large stars jostle around in each other's gravity fields, one or more of them might get flung out at high speed and end up far from their birthplace. This type of ejection can also happen when one star in a binary system goes supernova, releasing the other from its orbit and firing it across space in a straight line.

"Runaway star" Zeta Ophiuchi was ejected after its companion exploded in the Scorpius-Centaurus star-forming region around two million years ago. It is a hot star 20 times the mass of our Sun, 9 times the radius, and somewhat flattened, as material accreted from its pre-supernova companion made it spin more than 0.1 percent of the speed of light at its equator. Today, it sits alone at the center of a huge nebula called Sharpless 2-27.

You'll also see lots of material being ejected from its surface in a wind. As this wind expands and interacts with the dense clouds of Ophiuchus, their relative motion causes a giant wave of warm dust to pile up in front of the star, best seen with your goggles set to infrared.

The supernova that released Zeta Ophiuchi from its orbit is relatively close to Earth, just 350 light years away. Astronomers thought it might be one of the supernovae that "polluted" the surfaces of Earth and the Moon with tiny amounts of radioactive iron during the Pliocene and Miocene around three and seven million years ago, respectively, but it probably happened too recently.

Address Ophiuchus Dark Clouds, First Galactic Quadrant, The Milky Way (16h 37m 09.5s, -10° 34' 02") | **Getting there** Head towards the constellation of Ophiuchus, the serpent bearer, and travel for 390 years at the speed of light. | **Tip** B1706-16 is also worth a visit, as astronomers believe that it's most likely to be the pulsar left over when Zeta Ophiuchi's companion exploded (17h 09m 26.4s, -16° 40' 58").

DEEP SPACE

79 — Andromeda Galaxy
Our ever-closer neighbor

The Moon and planets in our night sky are all part of our local Solar System, while every individual star you can see is part of our Milky Way galaxy, most nearby. But even without picking up a pair of binoculars or a telescope, you can see much further. In the southern hemisphere, the Nubecula Major and Minor are satellite galaxies to our own, while in the north, there is a faint, fuzzy patch whose appearance belies its true grandeur. It is the Andromeda galaxy or Messier 31, and lies at a distance of around 24 million million million km (15 million million million mi) from Earth. It is one of the farthest objects visible to the naked eye, although only the core is bright enough to be seen – the whole galaxy spans a region on the sky four times bigger than the full Moon.

There's much more on display if you get closer. You'll see a large galaxy much like our own, with spiral arms of gas, dust, and star-forming regions wrapped around a central bulge of older, yellower stars, in the center of which is a supermassive black hole. It is also attended by a retinue of globular clusters and satellite galaxies. Astronomers are torn on whether the Andromeda galaxy has more stars than the Milky Way or fewer, and for more than half a century, they have debated whether or not the two galaxies have central bars. It's now believed that both indeed do.

But one thing is certain – the Andromeda galaxy is coming our way. Fortunately, it's not something you need to worry about during your visit. Even though the gap between Andromeda and the Milky Way is closing at around 400,000 km/h (250,000 mph), it'll be several billion years before they collide. When they do meet though, it'll be spectacular, gravity pulling out vast streams of stars as they fly through each other, gas and dust clouds smashing together to make new stars, before one day perhaps the two spirals merge to form a giant elliptical galaxy.

Address The Local Group, Deep Space (00h 42m 44.3s, +41° 16' 09") | **Getting there** Head towards the constellation of Andromeda, daughter of Cassiopeia, and travel for 2.5 million years at the speed of light. | **Tip** M32 is one of the Andromeda galaxy's satellites, a rather rare, compact elliptical galaxy. It's much smaller than its neighbor, and most of its stars are crowded in a dense core (00h 42m 41.8s, +40° 51' 55").

80 The Antennae
Magnificent chaos as galaxies collide

As you stare out at the cosmic carnage in front of you, it's perhaps hard to imagine that some 1.2 billion years ago, NGC4038 and NGC4039 were two ordinary, separate spiral galaxies, much like the Milky Way and Andromeda galaxies. But foreshadowing the future we face billions of years from now, NGC4038 and NGC4039 fell into a fatal attraction. Part of a larger group of at least 13 galaxies discovered in 1785 by William Herschel, their mutual gravity began to pull them together.

The distances between most stars is vast compared to their sizes, and so when galaxies collide, individual stars very rarely do. But gravity can influence the motions of stars, and as galaxies interact, they become distorted and may eject material. This can happen repeatedly as gravity brings the galaxies back together again and again. In the case of NGC4038 and NGC4039, two long streams of stars, gas, and dust have been drawn out across the sky, leading to the pair's nickname, the Antennae.

On the other hand, the giant clouds of gas and dust in galaxies *can* collide. They become compressed and start making new stars at prodigious rates. That's what you're seeing in the heart of the Antennae: a chaos of new star clusters surrounded by the telltale red glow of hydrogen ionized by the most massive, hot, blue stars. Hundreds of young globular clusters may also have been formed by the collision, while tendrils of dust twisting across the faces of the two galaxies and at the interaction zone between them show where yet more stars could form in the future. At the other end of the stellar life cycle, five supernovae have been seen in the Antennae over the past century.

Ultimately, the two galaxies will merge into one. In the meantime, during your visit to witness all of these wonders close-up, be careful not to get drawn into their ongoing, billion-year dance. You may never escape.

Address The NGC4038 Group, The Crater Cloud, Virgo Supercluster, Deep Space (12h 01m 54.0s, -18° 52' 43") | **Getting there** Head towards the constellation of Corvus, the crow, and travel for 45 million years at the speed of light. | **Tip** NGC4033 is in the same group, but it's an elliptical galaxy rather than spiral. It's interesting to see now, as the Antennae will likely turn into something similar in the future (12h 00m 34.7s, -17° 50' 33").

81 Arp 282
Caught in the act

One thing you will have noticed during your journeys is that galaxies are generally much closer to each other than the individual stars inside them are. For example, the distance between the Sun and our nearest neighbor, Proxima Centauri, is almost 30 million times the Sun's diameter. But the distance between the Milky Way and the Andromeda galaxy is only 15–25 times their size. And because everything is in motion, galaxies collide. In fact, they grow by interacting, ripping each other apart, and merging, especially if a big galaxy encounters a small one.

Arp 282 is a perfect example. On arrival, you'll first see a large galaxy called NGC169, a beautiful spiral with dust clouds tracing its arms and giving them a yellow-red glow. Soaring high above the plane of its disk is IC1559, a smaller, bluer spiral. Both have central black holes surrounded by accretion disks fizzing with energy at all wavelengths.

Between them is a bridge of gas, dust, and stars that has been pulled out by gravity. It's not immediately obvious which galaxy is the main source of this material – perhaps it's a mixture from both. On the other side of IC1559, however, is a blue, wedge-shaped tail that very likely comprises stars pulled out of the smaller galaxy by tidal forces. Look carefully at the larger NGC169 as well – the spiral arms on one side are more extended than on the other, perhaps drawn out as the smaller galaxy passed them earlier.

You can almost feel the pair moving, and if you had the patience to watch for millions of years, you'd witness them evolving. Maybe the smaller galaxy is moving away from the larger one and will continue to be pulled apart. If it isn't travelling fast enough to escape, it will fall back again to be swallowed by NGC169. And think again about the stars inside them. Even as the galaxies collide, their stars won't – they're too far apart.

Address Deep Space (00h 36m 51.6s, +23° 59' 28") | **Getting there** Head towards the constellation of Andromeda, daughter of Cassiopeia, and travel for 200–300 million years at the speed of light. | **Tip** NGC160 is in the same group but far enough away from the Arp 282 pair not to be interacting, at least at the moment. It is classified as a lenticular galaxy, but has a large, star-forming ring (00h 36m 04.1s, +23° 57' 28").

82 — Cartwheel Galaxy
Intergalactic hit and run

Left to their own devices and gravity, many larger galaxies adopt fairly simple, symmetric shapes. Spirals have a flattened disk and arms rotating around a bulge, perhaps with a bar across the middle, while ellipticals take on a more three-dimensional shape, from prolate to oblate or even triaxial. But when galaxies pass close to each other, things change dramatically, as tidal forces start pulling long tails, arcs, and shells of stars out of them. Something special happens, though, when one galaxy makes a bullseye hit right through the middle of another, as you'll see when you visit the Cartwheel Galaxy.

Discovered in 1941 by Fritz Zwicky (1898–1974), and also known as ESO 350-40, you'll immediately see why the Cartwheel is called a ring galaxy. In fact, it has two rings: a smaller inner one around the nucleus and a much larger outer one connected to it by a number of "spokes," the distorted vestiges of the galaxy's spiral arms. With your goggles set to infrared, you'll see how busy the outer ring and spokes are. Blue is where hot, young stars are being born, while red shows dust heated by them. All of this activity was kickstarted several hundred million years ago when another galaxy passed through the center of the Cartwheel. Like ripples spreading across a pond from a rock dropped into it, the changing gravitational field pushed gas and dust out from the center. As this material was compressed in shocks, it started to make huge numbers of new stars. There are many x-ray sources in the ring too, thanks to black holes left over when massive stars died.

The question is, what happened to the "bullet"? There are two smaller galaxies close to the Cartwheel, but it's probably neither of them. Rather, if you look a bit further afield, you'll find another spiral galaxy nearby known as G3 that's dragging a tail of neutral hydrogen gas behind. This tail is connected to the Cartwheel, marking G3 as the likely culprit.

Address Cartwheel Galaxy Group, Deep Space (00h 37m 41.1s, -33° 42' 58") | **Getting there** Head towards the constellation of Sculptor for 500 million years at the speed of light. | **Tip** Early into your long journey to the Cartwheel, you'll pass reasonably close to NGC134, a large spiral galaxy. Stop and look for its warped disk, likely the consequence of a much less dramatic galactic interaction (00h 30m 22.0s, -33° 14' 38").

83 Cigar Galaxy
Bursting with new stars

If you crave excitement, then the Cigar Galaxy (Messier 82, or M82) is the ideal destination. It has everything, including pulsars, black holes, gamma-ray bursts, effervescent star formation in its inner regions, exploding supernovae, and great gouts of gas spewing from its heart. It's quite the celestial fireworks show when seen close up.

Johann Elert Bode discovered M82 in 1774, along with its giant neighbor M81. The two galaxies orbit each other and periodically interact. A few hundred million years ago, the gravity of the larger M81 created tides in M82, channeling large quantities of gas and dust towards a small region around its central black hole. This high density of material led to the birth of hundreds of huge star clusters, making M82 a "starburst galaxy," far more luminous than our Milky Way. In turn, the most massive stars in those clusters only live a few million years before they explode as supernovae, cumulatively driving a colossal wind of high-speed gas out from the center.

Several of these supernovae have been observed in the past 20 years, most notably one called SN2014J, discovered in 2014 by a class of undergraduates using the University College London's small observatory. It turned out to be an exploding white dwarf star of a special kind known as a Type Ia supernova, the nearest to Earth in more than 40 years.

By the time you get there, SN2014J will long since have faded, but supernovae happen every 10 years or so in M82, so there's always a chance you'll see one. While waiting, you'll be able to marvel at the chaotic filigree structure of the red, hydrogen superwind rising high above the otherwise-normal-looking galaxy with its dusty spiral lanes. And deep inside the galaxy, there are other puzzles to be solved, including the origin of mysterious radio signals first heard in 2010. Most likely not aliens, but…

Address The M81 Group, Deep Space (09h 55m 51.7s, +69° 40' 47") | **Getting there** Head towards the constellation of Ursa Major, the great bear, and continue for around 12 million years at the speed of light. | **Tip** Arp's Loop is a peculiar arc wrapped around M81. It includes young blue star clusters orphaned in intergalactic space, along a stream pulled out between M81 and M82 (09h 57m 34.0s, +69° 16' 38").

DEEP SPACE

84 — The CMB
Left-over glow from the Big Bang

Whether you're on Earth or in space, the Cosmic Microwave Background (CMB) is there, day and night, a constant faint glow in all directions. Tune your goggles to radio wavelengths to be enveloped by the ghostly echoes of the Big Bang.

How can we still see something that happened long ago? Light doesn't travel instantly, and the farther away you look, the further back in time you see. It takes 8.3 minutes for photons to traverse the 150 million km (93 million mi) from the Sun, so we see it as it was 8.3 minutes ago. Look far enough away, and you'll see light that has taken 13.8 billion years to reach us – from soon after the Big Bang.

What is the light that you're seeing? Immediately after the Big Bang, our Universe was compressed into a tiny volume at unimaginably high densities and temperatures. Photons of light constantly bounced off electrons, making everything a hot, glowing fog. As space expanded, the volume increased, and the temperature dropped. After about 380,000 years, the temperature fell below about 2,700°C (4,900°F), and something special happened. The electrons combined with protons to form atoms and stopped scattering the photons. The Universe became transparent. It's those photons that we see arriving all around us still in the CMB.

At the time, the light would have been visible to humans, glowing orange. But as space has continued to expand, the light has been stretched out or "redshifted," lowering its apparent temperature to -270.4°C (-454.8°F), just 2.7°C (4.9°F) above absolute zero, emitting at radio wavelengths.

The CMB's glow is almost completely smooth across the sky, but look very carefully, as NASA's WMAP and ESA's Planck satellites did, and you'll see bumps with tiny temperature differences at the parts-per-million level. In the young Universe, those were the seeds that later formed clusters of galaxies, each containing countless stars.

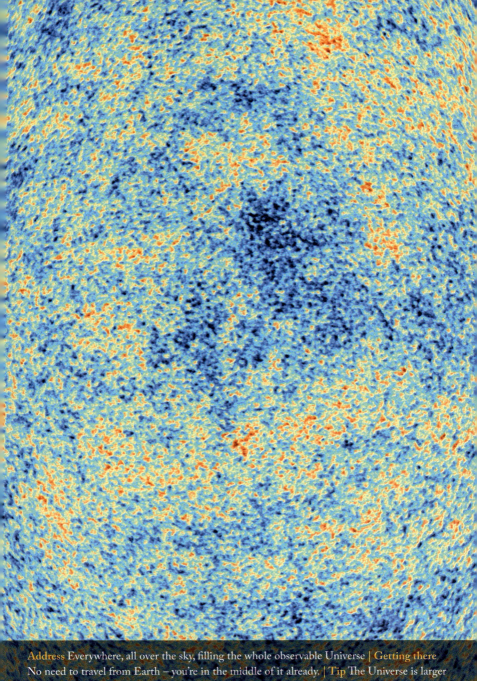

Address Everywhere, all over the sky, filling the whole observable Universe | **Getting there** No need to travel from Earth – you're in the middle of it already. | **Tip** The Universe is larger than the part we can see. As each day passes, the "wall" of the CMB recedes, and another light-day of the Universe is revealed. Who knows what mysteries lie beyond the wall?

DEEP SPACE

85 ESO 137-001
The pressure's on this high-speed medusa

Galaxies are fairly large compared to the space between them, and intergalactic collisions are common. These events can result in galaxies merging, one of the major ways that they evolve over cosmic time. But when galaxies collect in clusters, the space between them is not completely empty, which can also have a significant effect. You'll see this very clearly when you visit the Norma cluster near the center of the Great Attractor, and ESO 137-001, one of the best known "jellyfish galaxies."

Starting with the Norma cluster, your eyes are naturally drawn to the trillions of stars that shine at visible wavelengths. But tune your goggles to x-rays, and you'll see a glow from a plasma of mostly hydrogen and helium with small amounts of iron at 10–100 million °C (18–180 million °F) in the space between the galaxies. It's likely heated by jets from supermassive black holes and the energy released as galaxies merge. The plasma fills such a vast volume that it accounts for much more matter than the stars, gas, and dust in the galaxies.

Now turn your attention to ESO 137-001, a spiral galaxy barreling towards the core of the cluster at up to 10 million km/h (6.2 million mph). As it plows its way through the intracluster plasma, ram pressure strips gas from the galaxy to create a tail that's 260,000 light years long and emits x-rays, infrared, and visible light, resembling a jellyfish. Some of the material is found in downstream clumps that are now making stars. But as more gas is removed from the body of the galaxy itself, there's less available to make new stars there. Astronomers think that this process may be one of the main ways that spiral galaxies are turned into ellipticals, which have mostly stopped forming stars. So as long as you stay clear of the hot, stinging tail of the medusa during your visit, you'll learn a lot about how galaxies evolve.

Address Norma Galaxy Cluster, Deep Space (16h 13m 27.2s, -60° 45' 51") | **Getting there** Head towards the constellation of Triangulam Australe, the southern triangle, and travel for 220 million years at the speed of light. | **Tip** In the same direction but much closer to home in the Milky Way, the sparse, open star cluster NGC6052 is worth a stop for some quiet contemplation (16h 03m 18.5s, -60° 25' 36").

DEEP SPACE

86 ESO 306-17
The dangers of overconsumption

Galaxies naturally form in groups and clusters. Because of the relatively small distance between them, they interact, passing near each other and even colliding. Tidal forces can rip the galaxies apart, but more often than not, gravity pulls them together, and they merge to form ever larger ones. Slowly but surely, spirals grow and turn into lenticulars and then ellipticals, with the birth of new stars from gas and dust largely being quenched along the way. Giant ellipticals full of old stars often sit at the center of clusters, surrounded by smaller galaxies, which they continue to cannibalize.

So, what's the end game? You only have to travel to ESO 306-17 to find out. This "supergiant elliptical" is huge, and as the name suggests, one of the 10 largest known galaxies in the Universe. It is more than a million light years in diameter and its diffuse glow comes from perhaps a 100 trillion stars, making it more than ten times larger and up to 1,000 times more populous than our Milky Way. It truly is a giant.

ESO 306-17 is also a lonely giant. While you may see other galaxies around it, they're mostly in the background or foreground, not nearby. Don't feel sorry for it though – the reason it's isolated is that it has eaten almost all of its companions, the other galaxies in its former cluster. It's an example of a "fossil galaxy group." The only thing remaining of the original cluster is the giant, stuffed with the bones of all its victims.

But if you look carefully, you might see that ESO 306-17 is not fully alone. Buzzing around it are many globular clusters, small enough and with strong enough gravity holding them together to have been spared the depredations of its prodigious appetite. Astronomers are keen to discover whether this halo also contains any ultra-compact dwarf galaxies, stripped down to their dense cores by the giant but still managing to avoid being completely consumed.

Address Deep Space (05h 40m 06.7s, -40° 50' 11") | **Getting there** Head towards the constellation of Columba, the dove, and travel for 520 million years at the speed of light. | **Tip** ESO 383-76 is another supergiant elliptical and currently the largest known galaxy in the Universe at almost twice the size of ESO 306-17. But it is still surrounded by other galaxies in the Abell 3571 cluster and has the chance of growing even larger yet (13h 47m 28.4s, -32° 51' 54").

DEEP SPACE

87 Fornax A
Dusty heart of a hybrid galaxy

As you head towards the large cluster of more than 50 galaxies, you'll spot Fornax A on the outskirts. At first it looks fairly ordinary, a large ball of aging yellow stars with no sign of any spiral arms. As you get closer though, you'll see that its core is strewn with shredded red and brown dust clouds silhouetted against those stars. Fornax A, also known as NGC1316, is a lenticular galaxy, but an enigmatic one.

Spirals like our Milky Way have a flattened disk structure with arms full of gas, dust, and new stars wrapped around its central bulge. By contrast, elliptical galaxies are more three dimensional, perhaps spherical or rugby-ball shaped. Having exhausted almost all of their gas and dust, they show no active star formation. Lenticulars are thought to be an intermediate evolutionary stage between spirals and ellipticals. They have some features of both.

The turbulent history of Fornax A is written all over and around it. Towards the center are those chaotic dust clouds in arcs, filaments, and globules. It's thought that these are remnants of other galaxies it has swallowed in the last few billion years, with some of the dust blown back out from the nucleus. Look farther out to see shells, arcs, and ripples of stars surrounding it, gravitational echoes of those past merger events. Tune to radio wavelengths, and there are two giant lobes of hot plasma filled with relativistic electrons emitting energetically. Fornax A is one of the brightest radio sources in the sky.

To find the source of all that energy, you'll need to continue on to the core. There you'll discover a supermassive black hole, with around 140 million times the mass of our Sun, feeding actively on gas, dust, and stars from yet another galaxy that's being eaten. Strong magnetic fields around the black hole make jets that carry away excess material and inflate the giant radio lobes.

Address Fornax Cluster, Deep Space (03h 22m 41.7s, -37° 12' 29") | **Getting there** Head towards the constellation of Fornax, the furnace, and travel for 62 million years at the speed of light. | **Tip** NGC1317 is a smaller, barred spiral galaxy seen very close to Fornax A. It looks as though the two may be interacting, but NGC1317's distance is uncertain, and they may actually be far apart (03h 22m 44.3s, -37° 06' 13").

88 The Great Attractor
The inexorable pull of gravity

"The journey is just as important as the destination" is a saying that you might want to post on your dashboard as you set off towards this particular location, as it's not entirely certain what you'll find when you arrive. You won't be alone on your travels, though, because we're all being pulled in the same direction at 2.2 million km/h (1.4 million mph) by the gravity of a huge mass in space.

The Universe is in constant motion. Earth orbits the Sun, our Solar System orbits the Milky Way, and on the largest scales, galaxies are moving apart as spacetime, the fabric of the Universe, expands. But superimposed on that more-or-less uniform expansion are additional "peculiar" motions. If you measure the velocity of the Milky Way and our neighbors through the cosmos, you'll find that they're all moving in the direction of the constellations of Triangulam Australe and Norma.

What's pulling us that way? During the first part of your journey, the dense dust and star clouds in the inner regions of the Milky Way will obscure your view. But once you're beyond them, things will become clearer. You'll see a web of giant filaments and voids, made out of galaxies and groups of galaxies, the richest being the Norma cluster. Astronomers first thought that there was enough matter and gravity there to pull other clusters towards it, including our Local Group, so they dubbed it the "Great Attractor." But more recent measurements suggest that it's smaller than originally believed, and that sobriquet might better go to the Laniakea supercluster of perhaps 100,000 galaxies, which includes the Norma cluster at its core. However, much farther in the same direction is the larger Shapley supercluster, which might exert an even stronger pull on the Milky Way.

So once you've reached the original location of the Great Attractor, you might need to travel farther to solve this mystery.

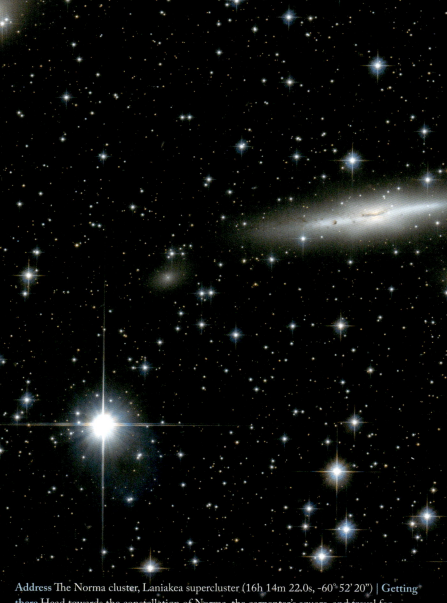

Address The Norma cluster, Laniakea supercluster (16h 14m 22.0s, -60° 52' 20") | **Getting there** Head towards the constellation of Norma, the carpenter's square, and travel for 225 million years at the speed of light. | **Tip** Menzel 3 is a planetary nebula in the northern part of Norma, about 8,000 light years away. It has a complex shape with a "head," "thorax," and "legs," the basis for its nickname the "Ant Nebula" (16h 17m 13.3s, -51° 59' 11").

DEEP SPACE

89 — Hanny's Voorwerp
The power of crowdsourcing

Many bodies in space have been known since antiquity, including most of our Solar System's planets. Others, like distant galaxies, were discovered by professional astronomers using large telescopes. However, given the vast volumes of space data available today, you too can find new and wondrous places from your own desktop.

Hanny's Voorwerp is named for Dutch teacher Hanny van Arkel, and "voorwerp" is Dutch for "object." Van Arkel was one of hundreds of thousands of citizen scientists who volunteered as part of the "Galaxy Zoo" project to classify over 900,000 galaxy images. When her computer showed her a picture of the galaxy IC2497 in Leo Minor to examine in 2007, she spotted something strange – neither she or the astronomers running the project knew what it was.

When you visit, you'll understand why. You'll first see the spiral galaxy, with its bright, dusty core. But next to it is a giant, green blob of gas: Hanny's Voorwerp. Green is an unusual color in space. Stars are typically blue, yellow, or red, depending on their temperature. While green can be found in nebulae, it's often mixed with reds and blues. So, why is this object so green? And why is it so far from the center of IC2497?

Following van Arkel's discovery, astronomers found that most of its light comes from ionized oxygen, which glows at green wavelengths. One theory is that a passing galaxy pulled a tail of gas out of IC2497, and some time later, the black hole at the galaxy's center turned on as a bright quasar. High-energy light was focused on part of the tail and ionized the oxygen. Even though the quasar has since turned off again, its photons are still hitting the gas, a kind of light echo. Similar objects collectively known as "Voorwerpjes," Dutch for "little objects," have been found elsewhere since, but they remain rare. So look out for more enigmatic green blobs during your travels.

Address Deep Space (09h 41m 03.8s, +34° 43' 37") | **Getting there** Head towards the constellation of Leo Minor, the smaller lion, and continue for around 650 million years at the speed of light. | **Tip** The Teacup Galaxy is a Voorwerpje discovered by the Galaxy Zoo team. It has a lovely loop of green, ionized gas resembling a handle on one side of the galaxy (14h 30m 29.9s, +13° 39' 12").

DEEP SPACE

90   JADES Origins Deep Field
The first galaxies in the Universe

Sometimes you just want to hop in your spacecraft and go. Pack your gear, pick a direction in the night sky that seems empty, and set off.

You'll first encounter stars in our own Milky Way, and later you will enter intergalactic space. Slow to a halt and amplify the view ahead of you. A few galaxies may loom bright and large, but between them you'll see many smaller and fainter ones, increasing in number as you collect more light. While it's true that galaxies come in many different types and sizes, the main effect here is one of distance – generally, the larger ones will be closer to you, the smaller ones farther away.

Because light travels at the finite speed of 299,792 km per second (186,262 mi per second), the farther away any given galaxy is, the longer it will have taken its light to reach you. As you look at the fainter, more distant ones, you're seeing them not as they are, but as they used to be, millions or billions of years ago. The final piece of the puzzle is that the Universe is only 13.8 billion years old, and so, at some great distance, you may spot the very first stars and galaxies that ever formed. This thought experiment is the motivation behind the "deep field" exposures made by pointing the Hubble and James Webb Space Telescopes for days, weeks, and even months at seemingly blank places on the sky.

The Universe is also expanding, which "redshifts," or stretches, the light of distant galaxies to longer wavelengths. Measuring a galaxy's redshift will give you a good idea of how far away it is and thus how far back in time you're looking. To date, the record holder is a tiny, faint speck called GS-z14-0, discovered in the JADES Origins Deep Field. Its redshift is 14.3, telling us we're seeing it less than 300 million years after the Big Bang. However, astronomers are still hoping to find even more distant, younger ones.

Address The Deepest of Deep Space (03h 32m 17.7s, -27° 51' 14") | **Getting there** Head towards the constellation of Fornax, the furnace. But no matter how fast or for how long you travel, you'll never reach those young galaxies because now they're old, and besides, the expansion of the Universe makes it impossible. | **Tip** The original Hubble Deep Field was observed over Christmas in 1995 and lies in another mostly empty piece of the sky in Ursa Major, the great bear (12h 36m 47.4s, +62° 13' 04").

DEEP SPACE

91 MACS J0025.4-1222
On the trail of the invisible

Most pictures of space are made by capturing light at many wavelengths. Sometimes the light comes directly from matter made of protons, neutrons, or electrons, whether in the form of hot stars, ionized gas, or warm dust. Other times, light is reflected or absorbed by compact objects like planets, or diffuse ones like clouds. All of this matter has mass, and its gravity determines how stars orbit around a galaxy. But the total mass of stars, gas, and dust in galaxies is not enough to explain the motions we measure. There must be much more material that has gravity, but that doesn't emit light or interact with it. This is "dark matter".

Visit MACS J0025.4-1222 to see how weird dark matter is. You'll come to a pair of huge galaxy clusters that collided a few hundred million years ago at speeds around 7.2 million km/h (4.5 million mph) and are now separating again. At first, you'll see the shining dots and clumps of light from the two clusters with their thousands of galaxies and trillions of stars. When they collided, the stars were too small to hit each other and glided past, but the majority of the normal matter in the galaxies was in gas, which did collide and heat up. Switch to x-ray vision: the hot gas glows pink between the two clusters, stuck there as the galaxies moved on.

Now it's time for a trick. If you look around the clusters, you'll find that their gravity slightly distorts images of other galaxies in the distant background. You can use this visual effect to map out the total mass in the two clusters, represented by blue. Unlike the gas, most of the mass has followed the clusters. Since most of that mass is dark, we can tell that dark matter doesn't collide with itself. So what is this strange stuff that has gravity but doesn't interact with light or with itself? We don't know, but many telescopes and satellites, including ESA's Euclid mission, as well as lab experiments, are on the case.

Address Deep Space (00h 25m 29.8s, -12° 22' 47") | **Getting there** Head towards the constellation of Cetus, the whale, and travel for 5.7 billion years at the speed of light. | **Tip** Worth a detour the same direction but rather closer is Arp 100, a pair of galaxies. IC18 is a dusty spiral with very extended tails pulled out of it, likely by IC19, an elliptical (00h 28m 35.0s, -11° 35' 12").

92 Messier 74
Design is how it works

Ask a child to draw a spiral galaxy, and she'll probably create something like Messier 74, or M74: face-on, with two clear, well-defined arms wrapped around the center. In fact, galaxies come in many other forms, and these so-called "grand design spirals" only account for about 10 percent of them. Still, it's well worth visiting M74 to get a clearer picture of how these majestic island universes work.

When you arrive, you may be a little underwhelmed at first. M74 is large, but it is also rather faint and can easily be overlooked, hence its nickname, the "Phantom Galaxy." However, the more light you can collect from it, the more beautiful it gets. In the center, you'll see a small bulge of old yellow stars. Fine tendrils of dust work their way outwards, marking the inner edges of the two prominent spiral arms while also extending across them. As the arms wind around the galaxy, they have the healthy glow of young stars, with blue clusters and red star-forming regions where the most massive stars are ionizing the hydrogen gas.

However, the stars aren't attached to the spiral arms and don't move with them. If you watched for millions of years, you'd see stars orbiting at different speeds depending on their distance from the center, while the spiral arms move as density waves around the galaxy at their own slower "pattern speed." An analogy is the way traffic bunches up when someone brakes suddenly, causing a high-density jam that moves slowly down the road, even though cars enter and leave the jam at higher speeds. In a galaxy, orbiting clouds of dust and gas enter the density wave, temporarily compressing them and leading to the birth of new stars. They then leave the other side of the wave and continue around the galaxy. The spiral waves need a trigger to get them going, like another small galaxy passing by, but once they start, they're self-sustaining.

Address M74 Group, Virgo Supercluster, Deep Space (01h 36m 41.7s, +15° 47' 01") | Getting there Head towards the constellation of Pisces, the fish, and travel for around 30 million years at the speed of light. | Tip NGC488 is one of many other galaxies worth visiting in Pisces. Unlike M74, it has lots of tightly-wound, evenly-spaced spiral arms (01h 21m 46.8s, +05° 15' 25").

93 Messier 87
King of its neighborhood

You're adrift in the middle of the Local Group of galaxies, including Andromeda and our own Milky Way. In the darkness, you spot a giant gathering of around 2,000 galaxies some 50 million light years away. That's the Virgo cluster, one of the largest in this part of the Universe, and its dominant member is the behemoth Messier 87 or M87, also known as NGC4486.

M87 was discovered by Charles Messier in 1781, and he classified it as a nebula based on its fuzzy, diffuse appearance. But once you get closer, you'll find that M87 is an elliptical galaxy made of trillions of stars in a three-dimensional ellipsoid, almost but not quite spherical. It has no flattened disk or spiral arms like our galaxy, and its shape is sustained by the looping orbits that its stars make in random directions around the center. Most of the stars are older, and there's little gas or dust from which new ones can be made, although some younger stars may have been added a billion years ago when M87 swallowed a neighbor. As perhaps befits its status as by far the largest galaxy in the Virgo cluster, M87 also has a vast army of attendants swarming around it, including some 100 dwarf galaxies and more than 10,000 globular clusters.

Beyond its sheer scale and grandeur, you'll also spot one of M87's most characteristic features, namely a blue jet of magnetized plasma spearing outwards from the galaxy's core towards intergalactic space. There are blobs of hot gas in the jet, and if you watch them closely, they'll seem to be moving faster than the speed of light. Don't worry though – it's just a trick of perspective, and physics isn't broken. The jet *is* moving at about half the speed of light, but it's also pointing close to the line of sight, which, combined with relativity, gives rise to the illusory superluminal motions. Where does the jet come from? A supermassive black hole called Pōwehi that lies at the center of M87. But that's another story.

Address Virgo Cluster, Deep Space (12h 30m 49.4s, +12° 23' 28") | **Getting there** Head towards the constellation of Virgo, the maiden, and go for around 54 million years at the speed of light. | **Tip** While M87 is the biggest galaxy in the Virgo Cluster, it's not the brightest. That honor goes to M49, which Messier also discovered a few years earlier. Visit to see the spectacle as it rips the nearby dwarf galaxy UGC7636 apart (12h 29m 46.7s, +08° 00' 01").

DEEP SPACE

94 Messier 106

Surveying the Universe

A good map needs an accurate scale to show how far you'll have to travel to reach your destination. It's surprisingly difficult to measure distances in space, and several different measurement techniques have been combined to create a "cosmic distance ladder."

For nearby stars, you can use the effect known as "parallax", watching how they appear to shift back and forth against the distant background during the year as Earth orbits the Sun. The bigger the wiggles a star makes, the closer it is. Simple trigonometry can then be used to calculate the distance. Farther away, the shifts become too small to measure, and other techniques are needed, including "standard candles." These are objects like Cepheids, RR Lyrae stars, and Type Ia supernovae that are assumed to have a known intrinsic luminosity. If you can measure their apparent brightness, you can calculate how far away they must be. But calibrating the brightness of these "candles" and connecting the various rungs of the distance ladder is tricky and error prone.

Messier 106, or NGC4258, is one of the fundamental calibrators of the distance ladder, making it a good place to visit. At first, it looks like a beautiful barred spiral galaxy with dusty arms filled with young stars and nebulae.

Tune your goggles to the radio, though, and focus on the core of the galaxy – it hosts a large black hole. You'll notice tiny, very bright spots in the gas spiraling around it, emitting at a particular wavelength linked to water vapor. The physical conditions in each spot have created a "megamaser," similar to a laser, but at microwave wavelengths. Using radio telescopes on Earth spread over 12,000 km (7,500 mi), astronomers watched the spots orbit the black hole. These motions could be turned into an accurate distance using trigonometry alone, and, as Messier 106 also contains Cepheid stars, this measurement has proved vital in calibrating the distance ladder.

Address Canes II Group, Deep Space (12h 18m 57.5s, +47° 18' 14") | **Getting there** Head towards the constellation of Canes Venatici, the hunting dogs, and travel for 24.7 million years at the speed of light. | **Tip** Also in Canes Venatici, but more than 400 times farther away than M106, is TON 618. It's a fascinating galaxy to visit, with a bright quasar in its heart powered by one of the largest black holes known, around 66 billion times the mass of our Sun (12h 28m 24.9s, +31° 28' 38").

95 NGC474

Shells on a galactic seashore

From a distance, NGC474 seems to be a fairly normal lenticular galaxy if you focus just on its bright and featureless central ball of stars. But turn up the gain on your cameras and look around the core for quite a bizarre sight. NGC474 is embedded in a halo of faint but sharply defined shells, extended streams, arcs, and whorls looping out from the center to vast distances.

Although strange in appearance, shell galaxies like NGC474 are relatively common, accounting for 10–20 percent of all ellipticals and lenticulars. They tend to be quite isolated or in small groups, as opposed to dense clusters, and NGC474 has just one close companion, a fairly normal looking spiral galaxy called NGC470. Together, they make up Arp 227, an entry in the famous *Atlas of Peculiar Galaxies* published in 1966 by Halton Arp (1927–2013).

The shells and streams around such galaxies are made of stars moved by tidal forces as they interact, but the exact mechanism has remained a mystery for decades. For example, has the gravity of the smaller NGC470 been pulling stars outwards from the larger NGC474 like ripples in a pond? Or is it the other way around, the larger galaxy pulling stars out of the smaller one? Another possibility is that NGC474 has been disrupting and merging with other smaller galaxies, sending their stars on strange new orbits around it. By comparing observations with supercomputer simulations, the current best explanation is the latter, as NGC474 merged with two smaller galaxies 1.3 and 0.9 billion years ago.

But NGC474 is also interacting with its neighbor NGC470 today, the two linked by streams of gas, perhaps fueling the intense burst of star formation seen in the nucleus of the smaller spiral. Come back in a few million years, and NGC470 may have been swallowed by NGC474, creating a new set of beautiful shells. That would definitely be something to see.

Address Deep Space (01h 20m 06.7s, +03° 24' 56") | **Getting there** Head towards the constellation of Pisces, the fishes, and travel for about 100 million years at the speed of light. | **Tip** The nearby NGC467 is another galaxy with shells and streams around it, but it's understudied by comparison. It's not clear whether NGC467 and NGC474 are connected, so it's worth a visit to check if you can (01h 19m 10.1s, +03° 18' 03").

DEEP SPACE

96 __ NGC660
What's your inclination?

In most spiral galaxies you'll encounter on your travels, the birth of new stars tends to be confined to the disk, in clouds of gas and dust in the arms that wrap around the central bulge. Sometimes you'll see interacting galaxies pulling vast bridges of material out between them, but rarely will you witness something quite as bizarre as NGC660. It's possible to make out dust lanes in its disk, and there is indeed star formation happening there. But very strangely, there is a huge ring canted at about 45°, soaring high above and below. This ring contains many red nebulae and hot, blue stars, also indicating ongoing star production.

NGC660 is a polar ring galaxy, a rare class characterized by gas, dust, and stars looping orthogonally over the poles. While that's not quite the case here, you may still be left wondering how that material has ended up in such an oddly inclined orbit around the galaxy. Or perhaps you've already correctly guessed that more than one galaxy has been involved.

The most likely explanation for the strange configuration of NGC660 is that more than a billion years ago, two spiral galaxies approached each other with their disks misaligned. They collided and merged. Material was then drawn from one into an inclined orbit around the other. Another option is that rather than a merger, material was drawn out of a second galaxy as it passed by. In either case, the violent encounter probably then triggered star formation in the ring, which is continuing to this day.

The longer you sit in deep space contemplating the weirdness of NGC660, the greater your chances of another surprise. At some point between 2008 and 2012, its core flared spectacularly at radio and x-ray wavelengths, probably due to a large gas cloud being swallowed by the central black hole. Maybe there's more yet to come from this most peculiar galaxy, and you'll be there to see it.

Address M74 Group, Virgo Supercluster, Deep Space (01h 43m 02.4s, +13° 38' 45") | **Getting there** Head towards the constellation of Pisces, the fishes, and travel for around 45 million years at the speed of light. | **Tip** NGC520, the Flying Ghost, is another beautiful destination in Pisces comprising a pair of galaxies colliding at an angle with strong tidal tails, perhaps similar to how NGC660 looked billions of years ago (01h 24m 35.4s, +03° 47' 30").

97 NGC1365

A galaxy walks into a bar …

The main reason to visit NGC1365 in the Fornax cluster is to see a classic example of a "barred spiral." Spiral galaxies comprise two main components: a central bulge of older stars and a number of arms spiraling outwards in which dense clouds of gas and dust form new stars. But roughly two-thirds of all spiral galaxies have an important, extra part, namely a linear bar of stars extending either side of the bulge. The main arms then usually start at the two ends of the bar. The Milky Way is a barred spiral, but it's hard to tell from the perspective of our home planet within it.

Discovered in 1826 by James Dunlop, NGC1365 has a bright nucleus crossed by brown tendrils of dust which also trace its east-west bar. At the ends of the bar, strong spiral arms curve away, one to the north, one to the south, traced by dust, young blue stars, and red star-forming regions. The shape of the arms gives NGC1365 its nickname the "Fornax Propellor Galaxy."

But as you get closer, you'll notice something a little strange. Embedded within the bar and closer to the nucleus is a second, smaller bar, rotated by about 60 degrees, making NGC1365 a double-barred spiral. Set your goggles to infrared to cut through the dust, and you'll also notice a collection of bright, young star clusters around the core, along with others where the main bar joins the spiral arms.

Bars are not rigid, rotating structures though. They're the result of density waves spreading out from the galaxy's nucleus and changing the orbits of stars around it from circular to elongated, leading to the bar shape. Similar processes channel material along the bar towards the middle, resulting in the birth of new stars and enhanced activity around the central black hole. Bars are not permanent either – it's thought that spiral galaxies change back and forth between being barred and unbarred every few billion years.

Address Fornax Cluster, Deep Space (03h 33m 36.4s, -36° 08' 25") | **Getting there** Head towards the constellation of Fornax, the furnace, and travel for 60 million years or so at the speed of light. | **Tip** The nearby NGC1369 is another barred galaxy in the Fornax cluster. It has been stripped of most of its gas and dust, so it's in the lenticular stage, perhaps en route to becoming an elliptical (03h 36m 45.3s, -36° 15' 23").

DEEP SPACE

98 NGC2276
Some galaxies have all the luck

For a stimulating vacation, try visiting the lovely, spiral NGC2276 – there's a lot going on here. On arrival, you'll immediately see that it appears distorted, its spiral arms more extended to one side of the nucleus than the other. One possible explanation for this asymmetry is encounters with other galaxies in the small group of which it's a part, in particular the eponymous NGC2300, a nearby elliptical galaxy. But that's probably not the main reason.

Look at the other side where things appear more squashed, and it's bursting with many regions of bright star formation. This compression is the result of the galaxy moving through hot gas in the cluster. Tune your goggles to the radio, and you'll indeed find that NGC2276 has a tail on the "stretched" side that extends for over 300,000 light years, more than five times the galaxy's width. That's a telltale sign of gas being stripped, thanks to ram pressure.

Turning to smaller scales, you'll discover just how active galaxy NGC2276 is. There have been six recorded supernovae in the past 60 years or so, a remarkably high number, mirroring the high rate of star formation. Look again at high energies, and you'll also find an unusual number of ultra-luminous x-ray sources. NGC2276 has at least eight, while our Milky Way has one at best. These are probably scaled-up versions of x-ray binary systems, where a neutron star or black hole pulls material from its stellar companion. This material gets heated to very high temperatures and then emits x-rays.

One x-ray source far from the center of the galaxy is especially rare and exciting. NGC2267-3c is associated with a black hole about 50,000 times the mass of our Sun, much larger than the black holes left over when large stars explode. Astronomers think these intermediate mass black holes may be the seeds from which the super-massive ones in the hearts of galaxies form.

Address NGC2300 Group, Deep Space (07h 27m 14.0s, +85° 45' 17") | **Getting there** Head towards the constellation of Cepheus, the mythical King of Aethiopia, and travel for 120 million years at the speed of light. | **Tip** Roughly a third of the way into your journey to NGC2276, visit NGC2146, a spectacularly messy spiral galaxy likely being distorted by an encounter. It is appropriately nicknamed the "Dusty Hand Galaxy" (06h 18m 37.6s, +78° 21' 24").

99 NGC2775
Pulling the wool over your eyes

Spiral galaxies come in many varieties, from the grand design spirals like Messier 74, with a few well-defined arms full of star-forming regions, to those with multiple arms, like NGC1232, and then those that have bars, like NGC1365. There are some real oddballs, too, like NGC2775, discovered in 1783 by William Herschel. When you reach this one, you'll be wondering, "Where exactly are the spiral arms?"

You can see the usual bulge of old, yellow stars in the middle of NGC2775. But it's far larger than normal, there's no hint of a bar, and there's little sign of any dust or cold molecular gas out of which new stars might be made. One theory is that the center of the galaxy was much more active in the past, making many massive stars, which then exploded as supernovae, combining to drive a mighty wind that pushed away any remaining gas and dust.

Farther out though, things start to get much more interesting. You'll see that the whole galaxy is wrapped tightly in a giant ring made of many small patches of blue, star-forming regions, with dust lanes marbling between and around them. An overall spiral pattern is discernible, but you'd be very hard pressed to pick out any specific continuous arms like their grand-design cousins have. This curdled or fluffy appearance is what gives galaxies like NGC2775 the name "flocculent spirals" after the Latin word *floccus,* a tuft of wool.

So what's going on in this galaxy? The best bet is that the outer ring is full of material that's making clusters of massive stars in various locations. They then start heating and compressing the gas around them, leading to a new round of cluster-making, a process astronomers call "self-propagating star formation." As all these star-forming regions develop, differential rotation in the galaxy shears them and pushes the associated dust out into the flocculent pattern you see before you.

Address Antlia-Hydra Cluster, Virgo Supercluster, Deep Space (09h 10m 20.1s, +07° 02' 17") | Getting there Head towards the constellation of Cancer, the crab, and continue for 67 million years at the speed of light. | Tip NGC2777 is a nearby, irregular galaxy, and there's a faint bridge connecting it to NGC2775. If your travel schedule allows, you can investigate to see if the two have interacted in the past (09h 10m 41.8s, +07° 12' 24").

DEEP SPACE

100 NGC4753
Cosmic filigree and shadow

The phrase "the beauty of physics" is used to describe how a rather small number of simple concepts, principles, constants, and equations combine to explain a vast array of phenomena, from microscopic to cosmic. It equally applies to some of the aesthetically pleasing outcomes of applying those rules, from the six-fold symmetry of snowflakes to the delicately etched, sinusoidal tracery of NGC4753.

Discovered by William Herschel in 1784, NGC4753 is a lenticular galaxy with a central disk of yellow stars and a large halo of younger, bluer ones. From afar, you'll see some reddish absorption across its face, which, as you get closer, resolves into a skein of dusty trails like a fine lace shawl draped around the galaxy. There's a compelling sense of geometry and motion as individual threads curve high above the midplane, reach a peak, and then curve down and under. It's as if some child-god has been playing with a vast, Spirograph toy.

In fact, it's just the beautiful outcome of physics. Around a billion years ago, NGC4753 collided at an angle with a smaller, gas-rich dwarf galaxy, triggering the birth of new stars. As the stars evolved, they belched out huge amounts of dust, which smeared out into a disk. Because the disk wasn't aligned with NGC4753's mid-plane, precession came into play. This is the effect you see when a spinning top slows, and its rotation axis starts to wobble. The difference here is that the disk of dust isn't solid, and parts closer to the nucleus precess faster than parts further away. Over time, this "differential precession" twisted the disk into a three-dimensional shape wrapped around NGC4753, its form traced by those dusty threads. Computer simulations reproduce this shape very well, though only by including lots of dark matter and its gravity. As we don't yet know what dark matter actually is, perhaps the physics here isn't quite so simple after all.

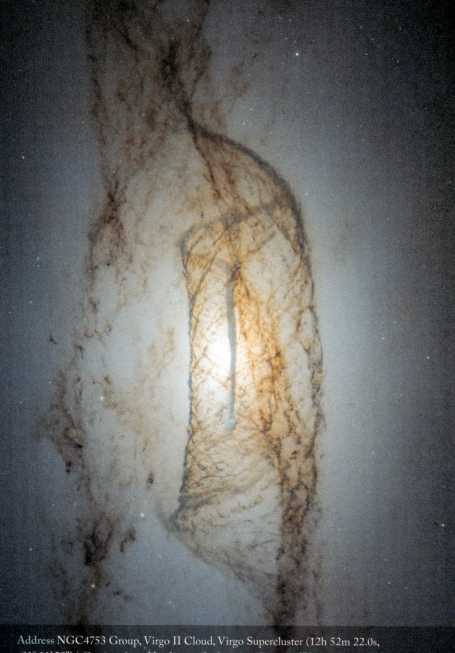

Address NGC4753 Group, Virgo II Cloud, Virgo Supercluster (12h 52m 22.0s, -01° 11' 58") | **Getting there** Head towards the constellation of Virgo, the maiden, and travel for 80 million years at the speed of light. | **Tip** NGC4643 is another lenticular galaxy in the same group as NGC4753 and is worth visiting to see its bright bulge and bar, as well as its smooth disk, free of star formation (12h 43m 20.1s, +01° 58' 42").

DEEP SPACE

101 NGC7331
The Milky Way's almost twin

As NGC7331 starts to loom large through the window of your spacecraft, you might be forgiven for thinking that you've accidentally flipped around during your journey and are heading home. Discovered by William Herschel in 1784, this galaxy looks similar in many ways to our own, with spiral arms around a central bulge and a supermassive black hole at its center. Sprinkled across the dusty arms are clusters of young, blue stars, perhaps only a few million years old, mingling with bright red patches of ionized hydrogen. NGC7331 is making stars at a similar rate as the Milky Way. But there will be subtle clues that you're indeed going the right way.

One difference is that the older, yellow stars in the bulge of NGC7331 rotate around the center of the galaxy in the opposite direction to those in its wider disk. That's not the case in the Milky Way. Zoom farther in, and you'll see that NGC7331's arms spiral all the way to the middle of the bulge like a pinwheel. Our Milky Way is different – it has a giant bar of older stars spanning the center, about 27,000 light years long, and our spiral arms only start at the end of the bar. Barred spirals are common in the Universe, but it wasn't until 2005 that astronomers using the NASA Spitzer infrared space telescope were able to confirm that the Milky Way is one of them. Turns out that it's quite hard to measure the structure of a galaxy when you're living inside it.

It's also worth keeping a keen eye on NGC7331 during your long journey there. Astronomers have seen three supernovae flaring and fading again in the galaxy over the past 70 years, the most recent of them in 2014. This latest massive star explosion was unusual. Some 127 days after the star's core had collapsed and then exploded, the supernova suddenly started showing hydrogen emission and brightened again at x-ray, infrared, and radio wavelengths as its shockwave collided with a surrounding torus of dense gas.

Address The Deer Lick Group, Deep Space (22h 37m 04.1s, +34° 24' 57") | **Getting there** Head towards the constellation of Pegasus, the flying horse, and continue for 40 million years at the speed of light. | **Tip** NGC7331 lies at the center of a group of galaxies, and the other four are sometimes referred to as the "fleas." They are unrelated, however, and lie another 250 to 320 million light years in the background. So it's not really a feasible side trip, but you'll have a nice view of them from here.

102 Nubecula Major
What's in a name?

With so much to see and do in our own bustling Milky Way galaxy, you may consider smaller satellite galaxies to be a bit provincial and not worth a visit. But you'd be mistaken. Visit the Nubecula Major to find a sparkling treasure chest teeming with interesting history and astronomical beauties, even if somewhat off the beaten track.

More commonly known as the Large Magellanic Cloud, it was named for the explorer Ferdinand Magellan (1480–1521), who saw the cloud and its smaller companion as patches of light in the night sky on his journey of circumnavigation, some years after other European and Arabian sailors had spotted them as well. But both "clouds" had, of course, long been known to Indigenous peoples of the Southern Hemisphere. To the Mapuche people of modern Chile and Argentina, they were the Rvganko, the water ponds. The Gamilaroi people of modern Australia see them as birds and linked to the afterlife. To recognise these older associations and Magellan's connection to a violent colonial legacy, many astronomers are calling for them to be renamed. Indeed, prior to the 1800s, they were known as the Nubecula Major and Minor, hence the usage here.

Despite its name, the Nubecula Major is a dwarf galaxy, just one percent the mass of the Milky Way. Its small size notwithstanding, it's full of gas and hosts many star-forming regions, including the Tarantula Nebula, which exceeds the largest in our galaxy. At the other end of the stellar life cycle, it's home to the most recent supernova visible to the naked eye, SN1987A. The Nubecula Major also has a giant bar of stars and a spiral arm, suggesting that it was once a full galaxy, now stripped of much of its material during an ongoing encounter with the Milky Way. Astronomers believe the two will merge in around 2.4 billion years' time, destroying the Nubecula Major and greatly changing the Milky Way.

Address Near the Milky Way, The Local Group, Deep Space (05h 23m 34.0s, -69° 45' 24") | **Getting there** Head towards the constellations of Dorado, the mahi-mahi, or Mensa, the table, for about 160,000 years at the speed of light. | **Tip** For a vertiginous experience, try crossing the faint bridge of gas and stars that connects the Nubecula Major and Minor, drawn out by tidal forces between the two (03h 11m, -73° 30').

103 Perseus Cluster
A life surfing the cosmic web

Young stars in the Milky Way are not uniformly distributed. Many are in clusters, which themselves are often strung out along the spiral arms of our galaxy like pearls on a necklace. On much larger scales, the same is true of galaxies. They also gather in clusters that are then part of superclusters. If you measure the positions and distances of millions of galaxies, you'll find that the majority lie on sheets and filaments of invisible dark matter, a mighty three-dimensional cosmic web. Visit the Perseus galaxy cluster, one of the most massive structures in the Universe, to understand how that web came to be and how it evolves over time.

You'll see thousands of individual galaxies, from dwarfs to giant ellipticals, many spread along an elongated central thread. The space between them is filled with a faint fuzz comprising vast numbers of stars ripped by tidal forces from their parent galaxies as they interacted and merged. That "intracluster light" follows the distribution of the dark matter, the gravity of which led to the formation of the filamentary structure of the cosmic web. The Perseus cluster is also permeated by a vast amount of superheated gas that becomes visible if you tune your goggles to x-rays. The black hole at the center of one of the most massive galaxies, NGC1275, is pumping ultralow-frequency sound waves into that plasma, far below human hearing.

Beyond the Perseus cluster, you'll also spot countless smaller, more distant galaxies. Studying their positions, distances, and small, gravity-induced changes in their shapes, astronomers can map out the three-dimensional distribution of dark matter. And by watching how the holes in the cosmic web change in size with time, they can measure the effects of dark energy, the mysterious force that is accelerating the expansion of the Universe. These are the goals of the Euclid Space Telescope, which has been surveying galaxies over large areas of the sky since 2023.

Address Perseus Cluster, Deep Space (03h 18m 55.5s, +41° 28' 12") | **Getting there** Head towards the constellation of Perseus, named for the mythical hero, and travel for 240 million years at the speed of light. | **Tip** On your way, you'll pass the famous variable star, Algol, also known as the Demon Star. It's a triple system where one of the stars passes in front of another every 2.7 days, partially eclipsing it and dimming the combined brightness by 70 percent for 10 hours (03h 08m 10.1s, +40° 57' 20").

DEEP SPACE

104_ Pōwehi

The dark heart of a supermassive black hole

As you plunge through the outer regions of the vast, elliptical galaxy that is Messier 87, the density of stars rises inexorably. Closer to the center, you'll see an asymmetric ring of extremely hot gas. In the middle of the ring lies … nothing. Or more precisely, nothing that you can see. This is a supermassive black hole with 6.5 billion times the mass of our Sun, one of the largest known. Inside it is the event horizon, where gravity becomes so strong that not even light can escape its pull.

The black hole at the center of M87 has the Hawaiian name 'Pōwehi,' meaning "the adorned fathomless dark creation." Using the Event Horizon Telescope in 2017, an array of radio dishes spanning Earth from Spain to Hawai'i, astronomers were able for the first time to achieve the incredible resolution needed to see the tiny ring of glowing plasma spiraling towards the spinning black hole, heated to very high temperatures before disappearing. The darkness you see inside the ring is effectively the shadow of the black hole against the bright plasma as its intense gravity distorts spacetime around it. Pōwehi swallows the equivalent of one sunlike star every decade or 90 Earth masses per day.

If you've grown tired of life in this Universe and want to take part in the ultimate physics experiment, venture closer and cross the event horizon inside the shadow. It will be surprisingly boring at first – Pōwehi is such a large black hole that its event horizon is several times the size of our Solar System, and tidal forces there are mild. But you're committed now, and within hours, things will become increasingly more gruesome as you fall towards the central singularity where the rapidly-changing gravitational field will stretch you into spaghetti. Your more sensible travel companions who remained outside will never see any of this, however. Thanks to relativity and time dilation, from their perspective, you'll be stuck on the event horizon forever, slowly fading away, a warning to future foolhardy tourists.

Address M87, Virgo Cluster, Deep Space (12h 30m 49.4s, +12° 23' 28") | **Getting there** Head towards the constellation of Virgo, the maiden, and travel for around 54 million years at the speed of light. | **Tip** HVGC-1 is a hypervelocity star cluster moving at more than 7.5 million km/h (4.7 million mph) away from M87. It may have been thrown out after a close encounter with Pōwehi (12h 30m 54.7s, +12° 40' 59").

DEEP SPACE

105 SMACS J0723.3-7327
Cosmic lens with a presidential seal of approval

Occasionally, arriving at a destination after a long journey can lead to disappointment. Some places are simply better viewed from a distance. And so it is with the cluster of galaxies known by its catalog and coordinate name SMACS J0723.3-7327, or SMACS 0723 for short. To be sure, the cluster is enormous, with perhaps 30 trillion stars and a total mass almost 600 times that of our Milky Way galaxy. More than enough going on to make it worth a visit, you might think. But what makes it truly exciting is what happens if you stay close to home and look at the cluster from near Earth.

The huge mass of the fuzzy white galaxies in the SMACS J0723 cluster distorts the spacetime around it, bending and focusing the light coming from the redder galaxies far beyond it. This distortion turns the cluster into a gravitational lens, one that can amplify and magnify images of those distant galaxies as seen from Earth's perspective. By leveraging this naturally occurring lens, astronomers can detect and resolve more detail in galaxies that would otherwise be too faint to be seen at all, including some that formed just a few hundred million years after the Big Bang.

Some of the galaxy images are bent into arcs around the SMACS 0723 cluster. Because the cluster's lens is lumpy and bumpy, not smooth, it's even possible to see multiple images of the same distant galaxies at different locations. Mapping these gravitational distortions and mirages allows astronomers to measure how normal matter and dark matter are distributed in the cluster.

These cosmic lenses can add a significant boost to mere human-made observatories. Thus it's no wonder that SMACS 0723 was one of the very first scientific targets of the James Webb Space Telescope, unveiled by then US President Joe Biden in July 2022. So on this occasion, relax in your armchair, and enjoy the majestic view across the whole Universe without going anywhere at all.

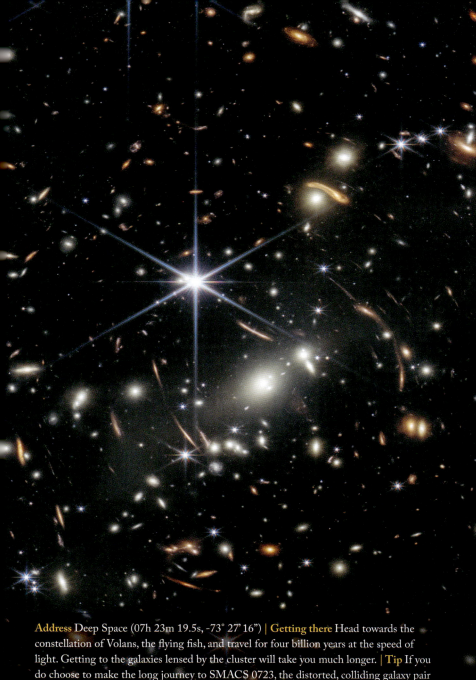

Address Deep Space (07h 23m 19.5s, -73° 27' 16") | **Getting there** Head towards the constellation of Volans, the flying fish, and travel for four billion years at the speed of light. Getting to the galaxies lensed by the cluster will take you much longer. | **Tip** If you do choose to make the long journey to SMACS 0723, the distorted, colliding galaxy pair NGC2442 and NGC2443, poetically named the Meathook, or the Cobra and Mouse, are worth a detour (07h 36m 23.8s, -69° 31' 51").

106 Sombrero Galaxy
A strange, dusty ring around a central monster

Galaxies come in all shapes and sizes, but sometimes appearances can be particularly deceptive. The galaxy now cataloged as Messier 104 and NGC4594 was discovered independently by both Pierre Méchain (1744–1804) and William Herschel in the late 18th century. It was seen to have a dark, dusty rim, giving it the appearance of a giant sun hat, hence its nickname, the Sombrero.

As you approach, you'll see a galaxy nearly edge-on, with star-forming arms wrapped around and connected to a central bulge. As a consequence, it was characterized as a spiral galaxy for many years. But if you set your googles to the infrared and look more closely, you'll see that most of the dust, gas, and young stars are confined to a broad, flattened ring. Inside that ring, however, the Sombrero has the properties of an elliptical galaxy, one which has exhausted its fuel supply and comprises only old stars. There are many globular star clusters orbiting in a halo around it, up to 2,000 perhaps, adding further weight to the idea that there's an elliptical galaxy in there.

So the Sombrero is neither a spiral nor an elliptical galaxy, but some combination of the two. It's tempting to think that two galaxies, one of each type, just crashed together, but then how did the ring keep its perfect shape? A more likely scenario is that a giant cloud of gas was drawn in by the gravity of an elliptical galaxy billions of years ago and ended up orbiting around it in a star-forming ring.

And as you get even closer, you'll find that its nucleus hosts a huge black hole. The speed at which stars are rotating around it tells astronomers that it is perhaps a billion times more massive than our Sun, the nearest black hole to Earth on that scale. Material falling towards the black hole is heated and emits in x-rays, the ultraviolet, and radio, causing the surrounding gas to become ionized and glow.

Address Virgo Supercluster, Deep Space (12h 39m 59.4s, -11° 37' 23") | **Getting there** Head towards the border between the constellations of Virgo, the maiden, and Corvus, the crow, and travel for 31 million years at the speed of light. | **Tip** The Sombrero is just one of hundreds of galaxies in the Virgo II Cloud, an archipelago 30 million light years long and the perfect starting point for some galaxy hopping (12 to 13.5h, -22° to +5°).

DEEP SPACE

107 Spanish Dancer Galaxy
I'm forever blowing bubbles

Taking a fresh look at familiar things can provide a better understanding of what's going on behind the scenes. At visible wavelengths, NGC1566 is a beautiful spiral galaxy, also known as the Spanish Dancer. But if you tune your goggles to infrared when you arrive, it will seem like you're looking at its skeleton.

Spiral galaxies comprise hundreds of billions of stars of all ages. Older, cooler ones tend to collect in the central bulge, while younger, hotter, bluer ones are more recently born out of gas and dust clouds in the spiral arms. Those clouds are often a mix of red, due to ionized hydrogen gas, and brown, as the dust absorbs the light of stars in and beyond them. This process heats the dust up, and if you view it at longer wavelengths, you can see it glow. Most stars are fainter in the infrared, so you can get a much clearer picture of a galaxy's underlying structure. In NGC1566, you'll see two well-delineated and symmetrical spiral arms with many regions of star formation in them. They're connected to a ring around the nucleus via a weak bar. The nucleus itself conceals a black hole with around 13 million times the mass of the Sun – it's active and has a hot accretion disk around it, which means the nucleus also shines brightly in the infrared.

You can also make out a large number of irregular bubbles and cavities in the spiral arms and trailing behind, ranging in size from a few light years across to thousands. The smaller ones are created by the strong winds and radiation from massive stars just before they explode as supernovae, while the largest ones result from the combined effect of winds and supernovae from many stars across a star-forming region. Astronomers are using the James Webb Space Telescope to look at the "bones" of NGC1566 and many other galaxies to understand how this "cosmic feedback" can affect their evolution.

Address The Dorado Group, Deep Space (04h 20m 00.4s, -54° 56' 17") | **Getting there** Head towards the constellation of Dorado, the mahi-mahi. The distance is poorly known, so expect to travel for at least 20 and perhaps up to 70 million years at the speed of light. | **Tip** NGC1515 is another beautiful spiral in the same group, but seen near edge-on. It has a dwarf companion, but don't be fooled by the lovely barred spiral nearby – it's actually much farther away (04h 04m 02.7s, -54° 06' 01").

108 Spindle Galaxy
O what a tangled web we weave

As you aim towards NGC5866 from Earth, it'll be obvious why it's also called the Spindle Galaxy. Seen edge-on, a line of dark dust clouds with frayed threads and filaments to either side is silhouetted against an unresolved haze of billions of stars in blue, with a subtle hint of yellow towards the center, where the bulge of older stars must be hiding.

Interesting mysteries remain about the identity of this galaxy. The first question is whether or not it appears in the famous catalog of Charles Messier. His collaborator Pierre Méchain spotted a faint object between the stars Omicron Boötis and Iota Draconis in 1781, and Messier catalogued it as M102. However, there were no coordinates in the printed list, and Méchain later disavowed his observation, saying that he had mistakenly "rediscovered" the nearby M101, the Pinwheel Galaxy. However, based on Méchain's original description and a hand-written position in Messier's notes, astronomers have debated for centuries whether M102 is just a duplicate or, in fact, a separate galaxy. If the latter, the best candidate is NGC5866, later independently discovered by William Herschel.

The second question is how to classify NGC5866. It's usually listed as a lenticular galaxy, halfway between a spiral and elliptical. As with Fornax A and NGC474, for example, lenticular galaxies shouldn't have much dust, or at least only near the nucleus. But as your eyes tell you, NGC5866 has plenty of dust all along the plane. So perhaps when you get there, you'll discover that it's actually a spiral making hot, young stars that tinge it blue.

Other things to look out for are the slight warp in the dusty disk, perhaps indicating a close encounter with another galaxy deep in the past, and the 100 or more globular clusters – their motions can be used to estimate the amount of dark matter surrounding NGC5866.

Address The NGC5866 Group, Deep Space (15h 06m 29.5s, +55° 45' 48") | **Getting there** Head towards the constellation of Draco, the dragon, and travel for 40–50 million years at the speed of light. | **Tip** NGC5907, aka the Knife-Edge or Splinter Galaxy, is a member of the same group and also seen near edge-on. The giant loops around it are a stream of stars dislodged as it swallowed a smaller galaxy (15h 15m 53.5s, +56° 19' 43").

DEEP SPACE

109 Stephan's Quintet
All is not what it seems in this group of galaxies

Fancy a vacation on a cosmic archipelago? Stephan's Quintet may be just the place. But beware: things are not quite as they appear, and the waters around the islands can get quite turbulent.

In 1920, an intense public discussion known as "The Great Debate" took place in Washington, DC between astronomers Harlow Shapley (1885–1972) and Heber Curtis (1872–1942). Shapley claimed that the so-called "spiral nebulae," such as Andromeda, were small and located within the vicinity of our Milky Way, which he believed comprised the entirety of the Universe. Curtis, on the other hand, said that they were "island universes," more distant, huge stellar systems, other galaxies like our own. Today, we know that Curtis was right and that the Universe is full of galaxies. And similar to the island chains of Hawai'i and the Seychelles, they often cluster together in spectacular groups.

Édouard Stephan (1837–1923) discovered the quintet of galaxies named for him in 1877, and it was one of the first regions to be studied by the James Webb Space Telescope in 2022. You'll find that the four large yellow galaxies are part of a common group at the same distance. The top three are colliding, perhaps doomed to merge into one giant galaxy in the distant future. Vast streams of gas are being swept out into intergalactic space where they crash together and heat up. New stars are being born there, far from the galaxy cores. One of the galaxies harbors a supermassive black hole surrounded by swirling hot matter.

But the fifth galaxy, the blue one on the left, is an interloper – it's in the foreground at perhaps only a quarter of the distance, seen by chance projected along the same line of sight. In fact, it's so close that you can see individual stars and star-forming regions in it. Be sure to take a look as you pass en route to the more distant galaxies with which it shares an archipelagic name.

Address Deep Space (22h 35m 57.5s, +33° 57' 36") | **Getting there** Head towards the constellation of Pegasus, the winged horse. Travelling at the speed of light, you'll pass NGC7320 after 40 million years and arrive at NGC7319, NGC7318a & b, and NGC7317 after about 290 million years. | **Tip** Although only four members of Stephan's Quintet are physically associated with each other, there is a fifth companion, NGC7320c. It's smaller and farther from the main group, so it's a better location for a quiet holiday.

DEEP SPACE

110 — Supernova 1987A
The most recent supernova in our neighborhood

Stars explode as supernovae in the Milky Way a few times per century, but only five appear in historical records spanning the past 1,000 years, and it's more than 400 years since the last one was seen by anyone on Earth. The problem is that most happen in the denser regions in the center of our galaxy, and distance and dust in the spiral arms conspire to make most of them invisible. You'll need to travel a little further afield to visit the scene of a more recent, naked-eye supernova, SN 1987A.

On 24 February 1987, Oscar Duhalde took an outdoor break from his work operating a telescope on Las Campanas in Chile. He noticed a new star in the Nubecula Major, a neighbor galaxy to the Milky Way. Elsewhere on the same mountain, astronomer Ian Shelton also saw it on a photographic plate he'd just taken. A short-lived, blue, supergiant star called Sanduleak -69 202, 20 times more massive than our Sun, had run out of nuclear fuel, collapsed, rebounded, and exploded. Light from the supernova, powered by the radioactive decay of nickel to cobalt and then to iron, reached Earth many millennia later and set off a frenzy of astronomical observations, from gamma-rays to radio waves and even neutrinos.

Visit today to see three rings that mark the site of the supernova. Sanduleak -69 202 had been ejecting material for thousands of years before it met its demise. A few years later, ultraviolet light from the explosion reached the rings and made them glow. Later, the slower blast wave caught up with the innermost ring, about a light year across, fragmenting it into a string of pearls and heating them to millions of degrees, hot enough to emit x-rays. While your eye may be drawn to those rings and the heavy elements forged in the star and during the explosion itself spewing into space, take a look at the center too, where the neutron star remnant has recently been discovered by astronomers using the James Webb Space Telescope.

Address Nubecula Major, Near the Milky Way, The Local Group, Deep Space (05h 35m 28.0s, -69° 16' 12") | **Getting there** Head towards the constellation of Dorado, the mahi-mahi, and travel for 168,000 years at the speed of light. | **Tip** Take a side trip to NGC2074, a nearby, colorful maelstrom of gas and dust betraying the presence of newly born stars, perhaps triggered by an earlier supernova like SN1987A (05h 39m 03.3s, -69° 29' 54").

111 Tarantula Nebula
Shelob's cosmic cousin

Our Milky Way galaxy is more than double the size of the neighboring Nubecula Major and contains up to 10 times as many stars. So you'd think that everything there would be on a smaller scale. But when you visit, you'll find that it's a hotbed of activity, and nowhere more so than in the Tarantula Nebula, the most prodigious region of star formation throughout the whole of the Local Group of more than 80 galaxies.

This nebula is vast and bright enough to be seen with the naked eye from Earth. As you get closer, though, you'll notice arcs, loops, and tendrils of gas and dust almost 1,000 light years across that some say resemble the legs of a giant spider – hence, its arachnoid name.

More formally known as 30 Doradus, the giant complex holds several star clusters. But it is primarily shaped and illuminated by the large open cluster NGC2070, home to more than 200 massive stars, many of which are binaries, and perhaps hundreds of thousands of smaller ones. Right at its core is a bright knot labelled R136a, which was originally thought to be a single star perhaps 1,000 times the mass of our Sun, defying all theoretical predictions for how large a star can be. However, progressively sharper images have revealed it to be a very dense group of eight hot, young stars with between 100–200 times the mass and millions of times the brightness of our Sun.

You'll be surprised not to see any evidence for supernova explosions or black holes in a region so full of massive stars. This lack suggests that the stars in the center of NGC2070 are just one to two million years old and haven't quite reached the end of their relatively brief lives yet. Astronomers think that the Tarantula Nebula may resemble the intense knots of star birth that can be seen in the most distant galaxies known. So you're visiting a laboratory for the study of how those galaxies were first made soon after the Big Bang.

Address Nubecula Major, Near the Milky Way, The Local Group, Deep Space (05h 38m 42.4s, -69° 06' 03") | **Getting there** Head towards the constellation of Doradus, the mahi-mahi, and travel for about 160,000 years at the speed of light. | **Tip** DEM L 190 is the brightest supernova remnant in the Nubecula Major, the remains of a massive star that exploded around 5,000 years ago. Its vividly colored shape gives it the nickname the "Brazil Nebula" (05h 26m 00.8s, -66° 04' 59").

Photo Credits

For links to the original sources for all images contained in this book, please visit: markmccaughrean.net/111-places-in-space

All astronomical coordinates are equinox J2000.0 and epoch 2000.0.

Carina Nebula (cover) / VLT HAWK-I, ESO, T. Preibisch, CC BY 4.0 INT;
Apollo 12 (ch. 1): Chandrayaan-2 / ISRO / image processing Mark McCaughrean, CC BY-SA;
Arrokoth (ch. 2): New Horizons / NASA, Johns Hopkins University Applied Physics Laboratory, Southwest Research Institute / image processing Roman Tkachenko;
Caloris Basin (ch. 3): MESSENGER / NASA, Johns Hopkins University Applied Physics Laboratory, Carnegie Institution of Washington;
Ceres (ch. 4): Dawn / NASA, JPL-Caltech, UCLA, MPS, DLR, IDA / image processing Daniel Macháček, CC BY-NC-ND;
Charon (ch. 5): New Horizons / NASA, Johns Hopkins University Applied Physics Laboratory, Southwest Research Institute;
The Cliffs of Hathor (ch. 6): Rosetta / ESA, OSIRIS Team / image processing Mark McCaughrean, CC BY-SA;
Comet 67P/Churyumov-Gerasimenko (ch. 7): Rosetta / ESA, OSIRIS Team / image processing Mark McCaughrean, CC BY-SA;
Dimorphos (ch. 8): DART / NASA, Johns Hopkins University Applied Physics Laboratory;
Dust Devils of Mars (ch. 9): Mars Reconnaissance Orbiter / NASA, JPL, University of Arizona / image processing Mark McCaughrean, CC BY-SA;
Europa (ch. 10): Galileo / NASA, JPL-Caltech, SETI Institute;
The "Face" on Mars (ch. 11): Mars Reconnaissance Orbiter / NASA, JPL, University of Arizona;
Far Side of the Moon (ch. 12): Apollo 16 / NASA, JSC, ASU / image processing Mark McCaughrean, CC BY-SA;
Ganymede (ch. 13): Juno / NASA, JPL-Caltech, SwRI, MSSS / image processing Kevin M. Gill, CC BY;
Hubble Space Telescope (ch. 14): Space Shuttle Atlantis / NASA;
Hyperion (ch. 15): Cassini / NASA, JPL-Caltech, Space Science Institute / image processing Phatom87;
International Space Station (ch. 16): Crew Dragon / ESA, NASA, Thomas Pesquet;

Io (ch. 17): Voyager 1 / NASA, JPL-Caltech / image processing Daniel Macháček;

Jupiter (ch. 18): Juno / NASA, JPL-Caltech, SwRI, MSSS / image processing Kevin M. Gill, CC BY;

Lutetia (ch. 19): Rosetta / ESA, OSIRIS Team / image processing Ted Stryk, CC BY-NC-ND;

Maat Mons (ch. 20): Magellan / NASA, USGS / image processing Mark McCaughrean, CC BY-SA;

Mars (ch. 21): Hope Mars Mission / UAE, EXI / image processing Jason Major, CC BY-NC-SA;

Mercury (ch. 22): MESSENGER / NASA, Johns Hopkins University Applied Physics Laboratory, Carnegie Institution of Washington;

Miranda (ch. 23): Voyager 2 / NASA, JPL-Caltech / image processing Jason Major, CC BY-NC-SA;

Near Side of the Moon (ch. 24): © Dennis Mellican;

Neptune (ch. 25): Voyager 2 / NASA, JPL-Caltech / image processing Ardenau4, CC0;

North Polar Hexagon (ch. 26): Cassini / NASA, JPL-Caltech, Space Science Institute / image processing Mark McCaughrean, CC BY-SA;

Olympus Mons (ch. 27): Mars Express / ESA, DLR, FU Berlin, HRSC / image processing Andrea Luck, CC BY-2.0;

'Oumuamua (ch. 28): ESO, Martin Kornmesser, Mark McCaughrean, CC-BY;

The Pale Blue Dot (ch. 29): International Space Station / ESA, NASA, Alex Gerst;

Phobos (ch. 30): Mars Express / ESA, DLR, FU Berlin, HRSC / image processing Andrea Luck, CC BY;

Plumes of Enceladus (ch. 31): Cassini / NASA, JPL-Caltech, Space Science Institute / image processing Mark McCaughrean, CC BY-SA;

Pluto (ch. 32): New Horizons / NASA, Johns Hopkins University Applied Physics Laboratory, Southwest Research Institute;

Rings of Saturn (ch. 33): Cassini / NASA, JPL-Caltech, Space Science Institute;

The Sun (ch. 34): Solar Orbiter / ESA, NASA, EUI Team / image processing Mark McCaughrean, CC BY-SA;

Sunspots (ch. 35): © Swedish 1-m Solar Telescope / L. Rouppe van der Voort (University of Oslo);

Titan (ch. 36): Cassini / NASA, JPL-Caltech, Space Science Institute / image processing Aster Cowart, CC BY-NC;

Triton (ch. 37): Voyager 2 / NASA, JPL-Caltech / image processing Ted Stryk;

Uranus (ch. 38): Voyager 2 / NASA, JPL-Caltech / image processing Ardenau4, CC0;

Venus (ch. 39): Akatsuki / JAXA, ISAS, DARTS / image processing Mark McCaughrean, CC BY-SA;

AG Carinae (ch. 40): Hubble Space Telescope / NASA, ESA, STScI, CC BY 4.0 INT;

Alpha Centauri System (ch. 41): © Nevenka Blagović Horvat & Miroslav Horvat;

Betelgeuse (ch. 42): Mark McCaughrean, CC BY-SA;

Boomerang Nebula (ch. 43): Hubble Space Telescope / NASA, ESA / image processing Judy Schmidt, CC BY 2.0;

Bubble Nebula (ch. 44): Hubble Space Telescope / NASA, ESA, & the Hubble Heritage Team, CC BY 4.0 INT;

Carina's Bok Globules (ch. 45): Hubble Space Telescope / NASA, ESA, N. Smith (University of California, Berkeley), the Hubble Heritage Team (STScI/AURA), & NOAO/AURA/NSF, CC BY 4.0 INT;

Cat's Eye Nebula (ch. 46): Hubble Space Telescope & Chandra / NASA, ESA, HEIC, & the Hubble Heritage Team (STScI/AURA); acknowledgment R. Corradi (Isaac Newton Group of Telescopes, Spain) & Z. Tsvetanov (NASA), CXC, RIT, J. Kastner et al., CC BY 4.0 INT;

Cederblad 110 (ch. 47): James Webb Space Telescope / NASA, ESA, CSA, M. Zamani (ESA/Webb), F. Sun, (Steward Observatory), Z. Smith (Open University), & the Ice Age ERS Team, CC BY 4.0 INT;

Cometary Globule 4 (ch. 48): DECam / CTIO, NOIRLab, DOE, NSF, AURA / image processing T. A. Rector (University of Alaska Anchorage/NSF NOIRLab), D. de Martin & M. Zamani (NSF NOIRLab), CC BY 4.0 INT;

Cosmic Bat Nebula (ch. 49): Ralph Wilhelm, Carpe-Noctem-Rooisand team, CC BY-ND;

Crab Nebula (ch. 50): James Webb Space Telescope / NASA, ESA, CSA, STScI, T. Temim, CC BY 4.0 INT;

CW Leonis (ch. 51): Hubble Space Telescope / NASA, ESA, Toshiya Ueta (University of Denver), Hyosun Kim (KASI);

Cygnus X (ch. 52): Spitzer & WISE / NASA, The Cygnus-X Spitzer Legacy Survey, WISE Image Service / image processing Judy Schmidt, CC BY 2.0;

Elephant's Trunk Nebula (ch. 53): Mayall 4m / NOIRLab / image processing T. A. Rector (University of Alaska Anchorage) & H. Schweiker (WIYN & NOIRLab/NSF/AURA), CC BY 4.0;

Galactic Center (ch. 54): MeerKAT, Spitzer, WISE / I. Heywood (SARAO), NASA, JPL-Caltech / image processing Judy Schmidt, CC BY;

HD209458b (ch. 55): Digitized Sky Survey / ESA, NASA, CC BY;

Herbig-Haro 212 (ch. 56): James Webb Space Telescope / NASA, ESA, CSA, Mark McCaughrean & Sam Pearson (ESA), CC BY-SA;

Horsehead Nebula (ch. 57): Euclid / ESA, Euclid Consortium, NASA / image processing J.-C. Cuillandre (CEA Paris-Saclay), G. Anselmi, CC BY-SA 3.0 IGO;

HR8799 (ch. 58): Large Binocular Telescope / A.-L. Maire;

Lagoon Nebula (ch. 59): Hubble Space Telescope / NASA, ESA, STScI, CC BY 4.0 INT;

LL Pegasi (ch. 60): Hubble Space Telescope / NASA, ESA, R. Sahai / image processing Judy Schmidt, CC BY;

The Milky Way (ch. 61): Gaia / ESA, DPAC / acknowledgment André Moitinho, CC BY-SA 3.0 IGO;

NGC1999 (ch. 62): Hubble Space Telescope & VST / ESA, NASA, ESO, K. Noll, CC BY 4.0 INT;

The OMC-1 Explosion (ch. 63): James Webb Space Telescope / NASA, ESA, CSA, Mark McCaughrean & Sam Pearson (ESA), CC BY-SA;

Omega Centauri (ch. 64): VLT Survey Telescope / ESO, CC BY;

Orion Nebula (ch. 65): Hubble Space Telescope / NASA, ESA, M. Robberto (Space Telescope Science Institute/ESA) & the Hubble Space Telescope Orion Treasury Project Team, CC BY 4.0 INT;

Pillars of Creation (ch. 66): James Webb Space Telescope / NASA, ESA, CSA, STScI / image processing Joseph DePasquale, Anton M. Koekemoer, Alyssa Pagan, CC BY 4.0 INT;

The Pleiades (ch. 67): Giuseppe Donatiello, CC0;

Polaris (ch. 68): © Miguel Claro | Dark Sky® Alqueva Official Observatory;

R Aquarii (ch. 69): Hubble Space Telescope / NASA, ESA, M. Stute, M. Karovska / image processing D. de Martin & M. Zamani (ESA/Hubble), CC BY 4.0 INT;

Ring Nebula (ch. 70): James Webb Space Telescope / NASA, ESA, CSA, STScI, CC BY 4.0 INT;

RS Puppis (ch. 71): Hubble Space Telescope / NASA, ESA, & the Hubble Heritage Team (STScI/AURA), Hubble/Europe Collaboration, acknowledgment: H. Bond (STScI & Penn State University), CC BY 4.0 INT;

Serpens Nebula (ch. 72): James Webb Space Telescope / NASA, ESA, CSA, STScI, Klaus Pontoppidan (NASA-JPL), Joel Green (STScI), CC BY 4.0 INT;

Sirius (ch. 73): Mark McCaughrean, CC BY-SA;

Taurus-Auriga Clouds (ch. 74): © Adam Block, Steward Observatory, University of Arizona;

Terzan 5 (ch. 75): Hubble Space Telescope / NASA, ESA, F. Ferraro, CC BY 4.0 INT;

Vela Supernova Remnant (ch. 76): DECam / CTIO, NOIRLab, DOE, NSF, AURA / image processing T. A. Rector (University of Alaska Anchorage/ NSF NOIRLab), M. Zamani & D. de Martin (NSF NOIRLab), CC BY 4.0 INT;

Westerlund 1 (ch. 77): James Webb Space Telescope / NASA, ESA, CSA, M. Zamani (ESA/Webb), M. G. Guarcello (INAF-OAPA) & the EWOCS team, CC BY 4.0 INT;

Zeta Ophiuchi (ch. 78): Spitzer Space Telescope / NASA, JPL-Caltech;

Andromeda Galaxy (ch. 79): Marcel Drechsler, Yann Sainty, CC BY-ND;

The Antennae (ch. 80): Hubble Space Telescope / NASA, ESA, & the Hubble Heritage Team (STScI/AURA)-ESA/Hubble Collaboration, acknowledgment B. Whitmore (STScI) & James Long (ESA/Hubble), CC BY 4.0 INT;

Arp 282 (ch. 81): Hubble Space Telescope / NASA, ESA, J. Dalcanton, Dark Energy Survey, DOE, FNAL, DECam, & CTIO, NOIRLab, NSF, AURA, SDSS, acknowledgment Judy Schmidt, CC BY 4.0 INT;

Cartwheel Galaxy (ch. 82): James Webb Space Telescope / NASA, ESA, CSA, STScI, CC BY 4.0 INT;

Cigar Galaxy (ch. 83): Hubble Space Telescope / NASA, ESA, & the Hubble Heritage Team (STScI/AURA), acknowledgment J. Gallagher (University of Wisconsin), M. Mountain (STScI), & P. Puxley (National Science Foundation), CC BY 4.0 INT;

The CMB (ch. 84): Planck / ESA, Planck Collaboration / image processing Mark McCaughrean;

ESO 137-001 (ch. 85): Hubble Space Telescope & Chandra X-ray Observatory / NASA, ESA, CXC, CC BY 4.0 INT;

ESO 306-17 (ch. 86): Hubble Space Telescope / NASA, ESA, & Michael West (ESO), CC BY 4.0 INT;

Fornax A (ch. 87): Hubble Space Telescope / NASA, ESA, & the Hubble Heritage Team (STScI, AURA), acknowledgment P. Goudfrooij (STScI), CC BY 4.0 INT;

The Great Attractor (ch. 88): Hubble Space Telescope / NASA, ESA, CC BY 4.0 INT;

Hanny's Voorwerp (ch. 89): Hubble Space Telescope / NASA, ESA, STScI, WIYN, W. Keel et al. / image processing Judy Schmidt, CC BY;

JADES Origins Deep Field (ch. 90): James Webb Space Telescope / NASA, ESA, CSA, J. Olmsted (STScI), S. Carniani (Scuola Normale Superiore), JADES Collaboration, CC BY 4.0 INT;

MACS J0025.4-1222 (ch. 91): Chandra X-Ray Observatory & Hubble Space Telescope / NASA, CXC, Stanford, S. Allen, NASA, STScI, UC Santa Barbara, M. Bradac, CC0;

Messier 74 (ch. 92): Hubble Space Telescope / NASA, ESA, R. Chandar, CC BY 4.0 INT;

Messier 87 (ch. 93): Hubble Space Telescope / NASA, ESA, & the Hubble Heritage Team (STScI/AURA), acknowledgment P. Cote (Herzberg Institute of Astrophysics) & E. Baltz (Stanford University), CC BY 4.0 INT;

Messier 106 (ch. 94): Hubble Space Telescope / NASA, ESA, the Hubble Heritage Team (STScI/AURA), & R. Gendler (for the Hubble Heritage Team), acknowledgment J. GaBany, CC BY 4.0 INT;

NGC474 (ch. 95): DECam / DES, DOE, Fermilab, NCSA & CTIO, NOIRLab, NSF, AURA / image processing DES, Jen Miller (Gemini Observatory/NSF NOIRLab), Travis Rector (University of Alaska Anchorage), Mahdi Zamani & Davide de Martin, CC BY 4.0;

NGC660 (ch. 96): International Gemini Observatory / AURA / image processing Travis Rector (University of Alaska Anchorage), CC BY;

NGC1365 (ch. 97): DECam / Dark Energy Survey, DOE, FNAL, CTIO, NOIRLab, NSF, AURA / image processing Travis Rector (University of Alaska Anchorage/NSF NOIRLab), Jen Miller (Gemini Observatory/NSF NOIRLab), Mahdi Zamani & Davide de Martin (NSF NOIRLab), CC BY;

NGC2276 (ch. 98): Hubble Space Telescope / NASA, ESA, P. Sell, acknowledgment L. Shatz, CC BY 4.0 INT;

NGC2775 (ch. 99): Hubble Space Telescope / NASA, ESA, J. Lee & the PHANGS-HST Team, acknowledgment Judy Schmidt, CC BY 4.0 INT;

NGC4753 (ch. 100): Hubble Space Telescope / NASA, ESA, L. Kelsey, CC BY 4.0 INT;

NGC7331 (ch. 101): Hubble Space Telescope / NASA, ESA, D. Milisavljevic (Purdue University), CC BY 4.0 INT;

Nubecula Major (ch. 102): © Robert Gendler;

Perseus Cluster (ch. 103): Euclid / ESA, Euclid Consortium, NASA / image processing J.-C. Cuillandre (CEA Paris-Saclay), G. Anselmi, CC BY-SA 3.0 IGO;

Pōwehi (ch. 104): Event Horizon Telescope / EHT Collaboration, L. Medeiros (Institute for Advanced Study), D. Psaltis (Georgia Tech), T. Lauer (NSF NOIRLab), & F. Ozel (Georgia Tech), CC BY;

SMACS J0723.3-7327 (ch. 105): James Webb Space Telescope / NASA, ESA, CSA, STScI, CC BY 4.0 INT;

Sombrero Galaxy (ch. 106): Hubble Space Telescope / NASA, ESA, & the Hubble Heritage Team (STScI/AURA), CC BY 4.0 INT;

Spanish Dancer Galaxy (ch. 107): James Webb Space Telescope / NASA, ESA, CSA / image processing Judy Schmidt, CC BY 2.0;

Spindle Galaxy (ch. 108): Hubble Space Telescope / NASA, ESA, & the Hubble Heritage Project (STScI, AURA), acknowledgment: William Keel (University of Alabama), CC BY 4.0 INT;

Stephan's Quintet (ch. 109): Hubble Space Telescope & James Webb Space Telescope / NASA, ESA & the Hubble SM4 ERO Team; NASA, ESA, CSA, & STScI / image processing Mark McCaughrean, CC BY-SA;

Supernova 1987A (ch. 110): Hubble Space Telescope / NASA, ESA, Robert P. Kirshner (CfA, Moore Foundation), Max Mutchler (STScI), Roberto Avila (STScI), CC BY 4.0 INT;

Tarantula Nebula (ch. 111): Hubble Space Telescope / NASA, ESA, E. Sabbi (STScI), CC BY 4.0 INT

Acknowledgments

"In space, no one can hear you scream," as the saying goes, and the lack of air to transmit sound waves and breathe is often accompanied by a lack of things to smell, taste, and touch, at least if you value your life. Thus, sight is the most important sense when it comes to exploring the Universe, and this book relies heavily on beautiful images made with telescopes and cameras on Earth, in space, and attached to our robots out in the Solar System.

Those systems are proposed, built, and operated by expert scientists, engineers, and technicians. Creating stunning images from raw data is also anything but "just taking a snap," requiring a lot of skill, patience, and aesthetic sense. Thus, I'd like to thank everyone involved at ESA, NASA, JAXA, and the many other national space agencies; at the big observatories run by ESO, NOIRLab and other organizations and international collaborations; and in industry and academia. Without them and the tax-payers around the world who fund the pursuit of scientific knowledge and understanding, this book could not exist.

Equally, a special mention is due to the many dedicated amateur astronomers who are doing amazing things with relatively small telescopes, and the "image processors" who download scientific data from open archives and create stunning views of the Universe as a hobby. In both cases, their work is often at or beyond the level of their professional colleagues, and I'm grateful to the generosity of those who have allowed me to share and showcase some of it here.

I am also grateful to those who study and provide information about space in scientific papers and books, and online in webpages, databases, and visualization tools. Wikipedia is a remarkable resource and fortunately free of any distortion or agenda when it comes to space.

On a more personal note, I'd like to give thanks to Bert Ulrich for suggesting me as a potential author for this book. To Karen Seiger for agreeing, for having the original idea, and for guiding me through the process as my editor. To Laura Olk and Ines Schmidtke for understanding that a book about space might involve different approaches to images and graphic design. To my scientific colleagues for answering my questions and for cutting me some slack while writing the book. To my friends, including Alex Milas, Karen O'Flaherty, and Louise Harra, for their support. And to my family, Sybille, Catriona, Finn, and Tigger for their love during what proved to be a challenging year.

Mark McCaughrean is an astronomer who has lived, worked, and taught in the UK, the USA, Germany, and The Netherlands, studying the birth of stars and planets. As the former Senior Advisor for Science & Exploration at the European Space Agency, he has worked on many space missions including Rosetta and the Hubble and James Webb Space Telescopes. A photographer and a cyclist, he is also the co-founder of Space Rocks, which celebrates space exploration and the art, music, and culture it inspires through public events and more.

www.markmccaughrean.net

The information in this book was accurate at the time of publication, but it can change at any time. Please confirm the details for the places you're planning to visit before you head out on your adventures.